THE ORIGIN AND EVOLUTION OF INTELLIGENCE

THE ORIGIN AND EVOLUTION OF INTELLIGENCE

EDITED BY

ARNOLD B. SCHEIBEL
J. WILLIAM SCHOPF

JONES AND BARTLETT PUBLISHERS
Sudbury, Massachusetts

Boston London Singapore

Editorial, Sales, and Customer Service Offices

Jones and Bartlett Publishers
40 Tall Pine Drive
Sudbury, MA 01776
(508) 443-5000
info@jbpub.com
http://www.jbpub.com

Jones and Bartlett Publishers International
Barb House, Barb Mews
London W6 7PA
UK

Library of Congress Cataloging-in-Publication Data

The origin and evolution of intelligence / edited by Arnold B. Scheibel, J. William Schopf.
p. cm.
Includes bibliographical references and index.
ISBN 0-7637-0365-6
1. Intelligence. 2. Animal intelligence. 3. Brain—Evolution.
I. Scheibel, Arnold B. II. Schopf, J. William.
QP398:075 1997
591.5'13--dc21 96-49305
CIP

Acquisitions Editor: Arthur C. Bartlett
Manufacturing Manager: Dana L. Cerrito
Editorial Production Service: Ocean Publication Services
Typesetting: Ruth Maassen
Cover Design: Hannus Design Associates
Printing and Binding: Courier Companies Inc.
Cover Printing: Coral Graphic Services, Inc.

A Contribution of the IGPP Center for the Study of Evolution and the Origin of Life (CSEOL), University of California, Los Angeles

Printed in the United States of America
01 00 99 98 97 10 9 8 7 6 5 4 3 2 1

CONTENTS

PREFACE

This volume is composed of a series of articles presented in March 1995 at the Eighth Annual Symposium of the UCLA Center for the Study of the Origin and Evolution of Life. A year earlier, in March 1994, the symposium topic was the Origin and Evolution of the Universe. The progression — from cosmic chaos to the rise of intelligence, from the Big Bang to the "big brain" — may not be inappropriate.

The authors of the contributions to this volume take as their canvas an enormous span of invertebrate and vertebrate life forms and wrestle with a vast array of problems ranging from direction finding in ants and birds to sociopolitical communication in monkeys, symbol manipulation in apes, and language use in humans. All these phenomena may be grouped under the general term intelligence, the unifying theme of the volume.

The contributors to this volume have studied the development and manifestations of intelligence in animals from insects to birds through monkeys to chimpanzees and humans. The latter, *Homo sapiens sapiens*, is a very recently evolved species, present on this planet for only an instant in geological time. Brief as this species' existence has been, much briefer has been the time during which it has begun to look at itself and its remarkable gifts with the tools of the biological, social, anthropological, psychological, and linguistic sciences.

We are only beginning the search for the origin and evolution of intelligence. Our quest may be early and faltering, but we must start somewhere . . .

Arnold B. Scheibel

BIOGRAPHICAL SKETCHES

Chapter 1: Prerational Intelligence: How Insects and Birds Find Their Way

RÜDIGER WEHNER

Rüdiger Wehner

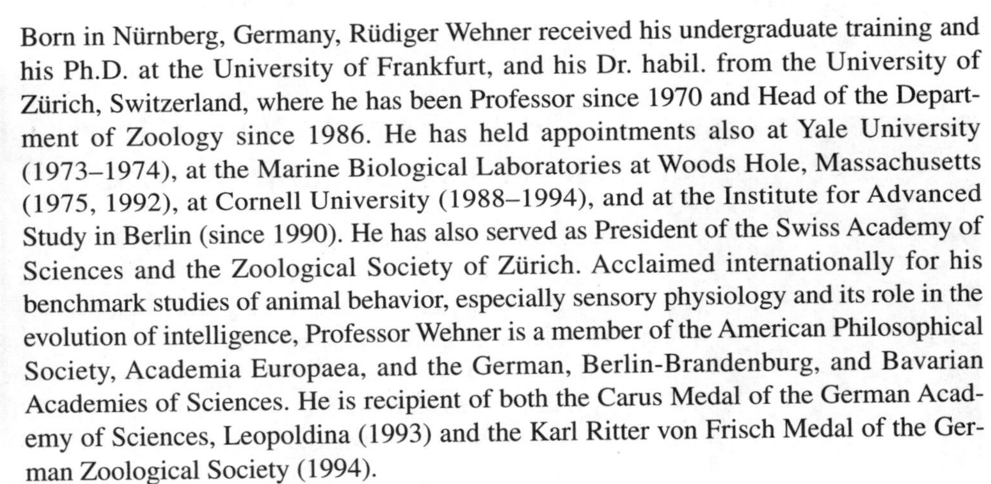

Born in Nürnberg, Germany, Rüdiger Wehner received his undergraduate training and his Ph.D. at the University of Frankfurt, and his Dr. habil. from the University of Zürich, Switzerland, where he has been Professor since 1970 and Head of the Department of Zoology since 1986. He has held appointments also at Yale University (1973–1974), at the Marine Biological Laboratories at Woods Hole, Massachusetts (1975, 1992), at Cornell University (1988–1994), and at the Institute for Advanced Study in Berlin (since 1990). He has also served as President of the Swiss Academy of Sciences and the Zoological Society of Zürich. Acclaimed internationally for his benchmark studies of animal behavior, especially sensory physiology and its role in the evolution of intelligence, Professor Wehner is a member of the American Philosophical Society, Academia Europaea, and the German, Berlin-Brandenburg, and Bavarian Academies of Sciences. He is recipient of both the Carus Medal of the German Academy of Sciences, Leopoldina (1993) and the Karl Ritter von Frisch Medal of the German Zoological Society (1994).

Chapter 2: Communication and the Minds of Monkeys

ROBERT M. SEYFARTH AND DOROTHY L. CHENEY

Robert M. Seyfarth

Robert M. Seyfarth and Dorothy L. Cheney, Professors of Psychology and Biology, respectively, at the University of Pennsylvania, comprise a world-class, husband-wife scientific team. Professors Seyfarth (B.A., Harvard University, 1970) and Cheney (B.A., Wellesley College, 1972) each received the Ph.D. degree from the University of Cambridge, England, in 1976–1977. They subsequently held appointments at Rockefeller University and UCLA. Both joined the University of Pennsylvania faculty in 1985. Together they have investigated the social behavior and vocalizations of diverse nonhuman primates in South Africa, Rwanda, Kenya, and Botswana. Joint recipients of the 1986 University of Pennsylvania Young Scientist Award, they coauthored *How Monkeys See the World* (Chicago: Univ. Chicago Press, 1990), a celebrated volume awarded the W. W. Howells Book Prize of the American Anthropological Association. In 1994, each was recipient of a prestigious Guggenheim Fellowship.

Chapter 3: Why Are We Afraid of Apes with Language?

SUE SAVAGE-RUMBAUGH

Sue Savage-Rumbaugh

A native of Springfield, Missouri, Sue Savage-Rumbaugh received her B.A. degree, *cum laude*, at Southwest Missouri State University and her advanced degrees (M.S., 1972; Ph.D., 1975) at the University of Oklahoma. She began her scientific career at the Yerkes Regional Primate Research Center at Emory University and in 1983 joined the faculty at Georgia State University in Atlanta; she is currently Professor of Biology and Psychology. A Fellow of the American Psychological Association and a National Lecturer for Sigma Xi, The National Research Society, she is author of three books, four films, and more than 100 scientific publications. Her research has focused on the cognitive processes and social and linguistic behavior of chimpanzees and is summarized in *Kanzi: The Ape at the Brink of the Human Mind* (New York: John Wiley, 1994), her highly readable and widely acclaimed most recent major work. A superb lecturer and accomplished scientist, Dr. Savage-Rumbaugh has carried out numerous studies in collaboration with her distinguished husband, Dr. Duane M. Rumbaugh, also a Professor of Psychology at Georgia State University.

Chapter 4: The Modular Nature of Human Intelligence

LEDA COSMIDES AND JOHN TOOBY

Leda Cosmides

John Tooby

Leda Cosmides and John Tooby, Professors of Psychology and Anthropology, respectively, at the University of California, Santa Barbara, are a highly effective wife-husband team, known widely for their leading work in the newly emerging field of Evolutionary Psychology. Both are graduates of Harvard University — Professor Cosmides in Biology (A.B., 1979) and Cognitive Psychology (Ph.D., 1985) and Professor Tooby in Psychology (A.B., 1975) and Biological Anthropology (Ph.D., 1989). Both have served as Fellows at the Center for Advanced Study in the Behavioral Sciences in Stanford, California.

Distinguished young scientists, both have received prestigious awards for their research contributions. In 1988, Dr. Cosmides was awarded the American Association for the Advancement of Science Prize for Behavioral Science Research and, in 1993, the American Psychological Association's Early Career Award. In 1991, Dr. Tooby was recipient of a Presidential Young Investigator Award from the U. S. National Science Foundation. Their recently published major work, *The Adapted Mind: Evolutionary Psychology and the Generation of Culture* (New York: Oxford Univ. Press, 1992), to which they contributed as coeditors and coauthors, is a benchmark volume in development of understanding of the evolutionary origins of human behavior.

Chapter 5: Evolution and Intelligence: Beyond the Argument from Design

TERRENCE W. DEACON

Terrence W. Deacon

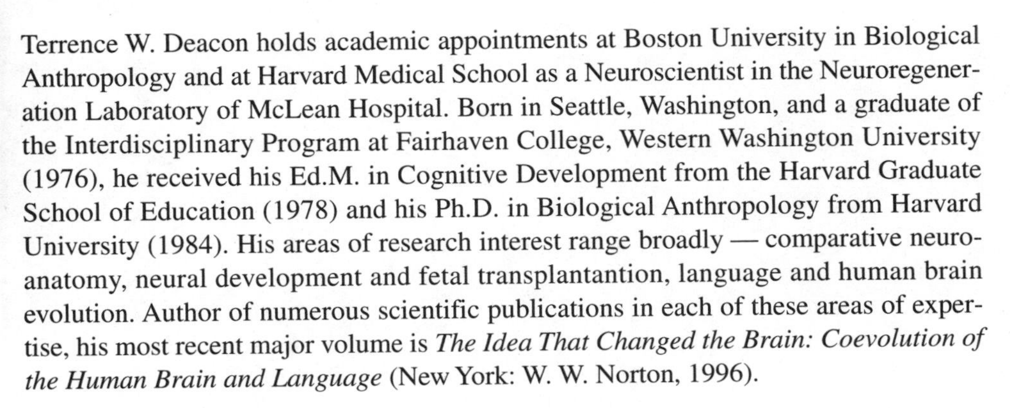

Terrence W. Deacon holds academic appointments at Boston University in Biological Anthropology and at Harvard Medical School as a Neuroscientist in the Neuroregeneration Laboratory of McLean Hospital. Born in Seattle, Washington, and a graduate of the Interdisciplinary Program at Fairhaven College, Western Washington University (1976), he received his Ed.M. in Cognitive Development from the Harvard Graduate School of Education (1978) and his Ph.D. in Biological Anthropology from Harvard University (1984). His areas of research interest range broadly — comparative neuroanatomy, neural development and fetal transplantantion, language and human brain evolution. Author of numerous scientific publications in each of these areas of expertise, his most recent major volume is *The Idea That Changed the Brain: Coevolution of the Human Brain and Language* (New York: W. W. Norton, 1996).

Chapter 6: Evolutionary Biology and the Evolution of Language

STEVEN PINKER

Steven Pinker

A native of Montreal, Canada, Steven Pinker is a graduate of McGill (B.A., 1976) and Harvard University (Ph.D., 1979). After teaching in the Departments of Psychology at Harvard and Stanford, in 1982 he joined the faculty of the Massachusetts Institute of Technology where he is Professor in the Department of Brain and Cognitive Sciences and Director of the McDonnell-Pew Center for Cognitive Science. Professor Pinker is recipient of the American Psychological Association's Award for an Early Career Contribution (1984), the MIT Graduate Teaching Award (1986), the McCandless Young Developmental Psychologist Award of the American Psychological Association (1986), and the Troland Research Award of the U. S. National Academy of Sciences (1993). He is a Fellow of the American Association for the Advancement of Science, the American Psychological Association, and the American Psychological Society. His most recent major work, *The Language Instinct* (New York: William Morrow, 1994), was awarded the William James Book Prize of the American Psychological Association and acclaimed by the *New York Times Book Review* as one of the Ten Best Books of 1994.

INTRODUCTION

Arnold B. Scheibel*

We live on a smallish inner planet circling an average star, one of perhaps a hundred billion stars making up a spiral galaxy that in turn is a member of a galaxy cluster, one of millions of clusters immersed in space-time. As far as we know, our Universe is endless, yet expanding, the sum of all the matter and energy that ever was and perhaps that ever shall be. But the inanimate Universe, despite the cataclysmic events that mark its history, remains inchoate and insensate, the product of massive singularities and stochastic events governed only by the laws of physics.

Assuming for the moment that our planet is the only island of life, it is intriguing to speculate that the onset of biological processes and ultimately, the appearance of human intelligence, may have represented a new and special phase of cosmic development, one in which the Universe finally became sensate and achieved the capability of contemplating itself. If so, the content of this volume stands in very special relation to, as a logical continuation of, the proceedings of the previous year's symposium. The titanic forces and gigantic stage on which the drama of the Universe has been played out dwarf anything we can compare it with. Yet, in some ways, the remarkable and statistically improbable development of the special biological faculties of sensation, motivation, voluntary behavior, and understanding already bid fair to match, in their ultimate scope, anything that has gone before.

Intelligence, as we know it, develops within the nexus of groups of living entities called cells. Originally, single-celled organisms such as bacteria and protozoans subsumed all biological functions, including primitive means of sensing and responding to the environment. With the appearance of multicellular organisms, groups of cellular entities became specialized to handle the sensory and motor needs of the organism. For hundreds of millions of years, the increasingly complex invertebrate forms fine-tuned the structure and function of their nerve cell aggregates, having such success that today invertebrates remain the most numerous, and arguably, the most successfully adapted animal life forms on Earth.

The typical invertebrate nervous system consists of a number of discrete cell clusters, called *ganglia*, bound together into communicating nets by fiber bundles called *connectives*. Each ganglion has a limited range of information-handling responsibilities and is located close to the body area over which it exerts jurisdiction. For example, in

*Departments of Neurobiology and Psychiatry, School of Medicine, University of California, Los Angeles.

the octopus, an extremely successful and highly intelligent cephalopod mollusc, some ganglia are specifically assigned to control of the arms, and others are concerned with function of the mouth parts, the eyes, and so forth. It is notable that the major ganglia are grouped more closely in the octopus than are ganglia in most invertebrates; in some cases, they adhere to each other — foretelling, in a sense, the makeup of the vertebrate brain. From this point on, in evolution, all "higher" organisms, the vertebrates, are equipped with a central nervous system made up of the spinal cord, the brain stem (the rostral continuation of the spinal cord), and a paired forebrain or telencephalon, made up largely of the cerebral hemispheres and a group of deeper nuclei, the basal ganglia.

As one ascends the vertebrate line, the most obvious nervous system change is the continuous size increase of the forebrain. It seems likely that in its most primitive form (for example, in amphioxus, a notochordal prevertebrate animal, and in cyclostome, an early type of vertebrate), the telencephalon was primarily an elaboration of upper brain-stem tissue generated to receive olfactory information. Olfaction was probably the earliest of the distance-interactive (teloreceptive) senses. It was important for a mobile animal to be aware of what lay ahead, before physical contact was actually established with the object. Was that object dangerous? Edible? A potential sexual partner? Such critical distinctions and the behavioral decisions to which they led demanded some type of record of previous encounters with which the data being processed could be compared. Clearly, such a system had to possess sorting and recognition mechanisms and some type of trace repository or memory. The primitive telencephalic structures (archi and paleocortex) that emerged to fill these olfactory needs undoubtedly became the model for other cerebral cortical systems that evolved as the sensory-motor repertoire of vertebrates became richer.

The continuous overgrowth of the cerebral cortex reached one climax in the appearance of folding of the cortical surface into swellings (gyri) and indentations (sulci) in more evolved mammalian vertebrates. We are reasonably certain that gyrencephalic cortices developed as a means of fitting more cortical area into a skull of limited volume. Rodents, such as rats and rabbits, have smooth (lissencephalic) cortices, but carnivores (cats and dogs, for example), primates, and humans have cortices with increasing degrees of gyration.

It is assumed that a larger cortical surface area — or, in different terms, a higher encephalization quotient (EQ) (Jerison, 1973) — correlates positively with more intelligent behavior. If so, the second climactic event in brain development occurred with the appearance of a series of hominid forms beginning approximately 5 to 4.5 million years ago and culminated in the emergence of modern man (*Homo sapiens sapiens*) within the last 60,000 to 40,000 years. Within a total period of less than 5 million years, the brain, primarily the cerebral hemispheres, tripled in size from approximately 450 cm^3 to 1200–1500 cm^3. No biological organ (except perhaps the horse's hoof) is known to have undergone so rapid an increase in size.

As remarkable as the rapid growth appears to be, research in recent decades has revealed mechanisms that may serve as substrate to this change. The brain maintains some degree of plasticity during the life of the organism and responds to environmental challenge (enrichment) with increased growth of connections (Diamond, 1976, 1988). In other words, increased computational complexity leads to enhanced structural complexity. The implications of this remarkable capability, both for the short and long term, are obvious.

A number of questions come to mind. What biological and environmental forces in prehominid and hominid history triggered the rapid growth? Is the human brain still growing as each generation is confronted by a more complex world? What are the implications of these remarkable changes for the future of humanity?

Intelligence is an elusive term. Most of us think we know what we (and others) mean when we use this term, yet defining it is a problem. Its Latin roots, *inter* and *ligere*, mean "to bind together within," suggesting perhaps the enormous synthesis of information that is the root of intelligent behavior. The *Oxford Universal Dictionary* (1953) lists at least six definitions, including "the faculty of understanding," "the action or fact of mentally apprehending something," and "understanding as a quality, admitting of degree." *Wiley's International Medical Dictionary* (1986) ranges more widely, with a series of suggestions: (1) "a quality of behavior demonstrating the degree to which an organism is capable to learn quickly and to adapt responses rapidly and effectively when faced with novel situations" and (2) "the ability to manipulate symbols and to grasp abstract relationships in problem-solving and to respond flexibly to changing demands within a given context. Human intelligence would encompass the collective repertory of cognitive powers possessed by an individual which can be brought to bear on the solution of difficult and complex problems, such as reason, insight, foresight, judgment, and imagination."

Wiley's suggests four types of intelligence: abstract, artificial, mechanical, and social, which are reminiscent of Gardner's (1993) somewhat more extensive seven categories of intelligence in the human child — verbal, mathematical-logical, spatial, kinesthetic, musical, personal-interpersonal, and intrapsychic — as well as those considered by Goleman (1995) as "emotional intelligence."

As I use the term, *intelligence* denotes a range of attributes that mold behavior and enable an organism to respond appropriately and successfully to the challenges of its environment. Intrinsic to this definition is the assumption that intelligence is many faceted. Successful adaptation to its own niche marks an animal as intelligent whether the physical mechanism is limited to the several hundred neurons of the tiny nematode worm *Caenorhabditis elegans* or gifted with a 1400-g brain like that of *Homo sapiens*.

REFERENCES

Diamond, M.C. 1976. Anatomical brain changes induced by environment. In: McGaugh, J.D. and Petinovitch, I. (Eds.), *Knowing, Thinking and Believing* (New York: Plenum Press), pp. 215–241.

Diamond, M.C. 1988. *Enriching Heredity* (New York: The Free Press), 191 pp.

Gardner, H. 1993. *Multiple Intelligence, The Theory in Practice* (New York: Basic Books), 304 pp.

Goleman, D. 1995. *Emotional Intelligence* (New York: Bantam Books), 352 pp.

Landau, S. (Ed.). 1986. *Wiley's International Dictionary of Medicine and Biology* (New York: Wiley), 3200 pp.

Onions, C. (Ed.). 1955. *The Oxford University Dictionary* (3rd ed.) (Oxford: Clarendon Press), 2515 pp.

Jerison, H. 1973. *Evolution of the Brain and Intelligence* (New York: Academic Press), 482 pp.

CHAPTER 1

PRERATIONAL INTELLIGENCE — HOW INSECTS AND BIRDS FIND THEIR WAY

■

Rüdiger Wehner*

"Say from whence you owe the strange intelligence." This is the question Macbeth asks when the witches foretell that he will become the father of kings but not a king himself. Here, as always in Shakespeare, *intelligence* means news, or the way of acquiring news. To a biologist, and especially to an evolutionary biologist, this is a comfortably broad definition; and, because a universally accepted more restricted definition of intelligence has not yet emerged in any scientific literature, I will stay with this one. Do the witches, Macbeth's ghostly military intelligence, behave in a *rational* way as they communicate their news to Macbeth, and what can be said about his rationality in coping with the news? Whatever the answers to these questions and however philosophers, anthropologists, and men of letters have ever tried to define terms such as intelligence and rationality, the definitions have always been biased by the tacit assumption that we humans are unique, that we are uniquely endowed with intelligence and thus set apart from beast and birds, from bees and bugs — that is, from organisms that are lumped together under the term *animals* or, as Descartes said, "thoughtless brutes."

However, because humans have evolved, not suddenly sprung into existence, intelligence has also evolved. Because of the multitude of evolutionary lines, intelligent behaviors have come in various guises. My tenet is that by studying certain different prerationally intelligent behaviors we may come across some general aspects of how brains deal with complex tasks, and — perhaps more important — how they do not. From study we can learn about the evolutionary routes that have led to human intelligence and the basic structure of the human mind.

At this juncture I want to avoid the difficulties of explaining in detail what I mean by prerational but intelligent behavior. As always, it is devilishly difficult and potentially misleading to define the apparently obvious in abstract terms — once expressed quite concisely by Louis Armstrong when asked to define jazz: "Man, if you gotta ask, you'll

*Department of Zoology, University of Zürich, Switzerland CH-8057

never know." Hence, for the time being, let prerational intelligence mean whatever you think it means, and let me introduce you to a part of the world that is more tangible, namely, the salt pans of the Sahara Desert, one of the most hostile habitats present on the surface of our planet.

> Most people actually laugh at me for carrying on research in these matters, and I am accused of busying myself with trifles. It is, however, a great comfort to me in my vast toil to know that Nature, too, not I alone, incurs this concept.
>
> Pliny the Elder, *Natural History*, circa 50 A.D.

SMALL-SCALE NAVIGATION — THE INSECT FORAGER

The salt pans of the Sahara Desert — vast expanses of flat, hot, and dry terrain — are inhabited by very few animals. *Cataglyphis fortis*, a skillful and vivacious species of ant, is certainly the most remarkable of these species. It dashes, leaps, and scrambles across the desert surface, and sweeps it for widely scattered food particles, mostly carcasses of other insects that have succumbed to the stress of the harsh desert environment. In searching, the ant leaves its underground colony and scavenges across the desert floor for distances of more than 200 m, winding its way in a tortuous search for fodder. Once it has found a bit of food, the ant grasps it, turns around, and runs more or less directly back to the starting point of its foraging excursion, a tiny hole that leads to its subterranean nest. The ant does not follow a pheromone trail nor retrace its outbound path by other means. Instead, it integrates its path and then moves along its line over new territory (Figure 1.1). How does the ant accomplish this feat of computation?

Before we tackle this particular question, permit me to digress for the sake of perspective and ask why intelligence and its material substrate, the nervous system, evolved in the first place. The answer is that in early Cambrian or even Precambrian times (550 million years ago or earlier) certain groups of organisms became heterotrophic and dependent on particulate food. In the **autotrophic** plant world — that is, in the world of cyanobacteria, unicellular algae, and higher plants — nutrients are ubiquitous and are distributed more or less homogeneously. For **heterotrophic** organisms, the "eaters" of the world, nutrient material is spread about in patches and is inhomogeneous and unpredictable in both space and time. To survive, the early heterotrophic organisms must have been able to move and to direct their movement toward potential food in efficient ways. We call the forms of life that accomplish these tasks *animals* and their sensorimotor steering devices *nervous systems*. The inhomogeneity and unpredictability of the external world and the need to cope with the environmental contingencies opened the floodgates for the evolution of animal behavior and intelligence.

Let me now return to my favorite heterotrophic organism, *Cataglyphis*, and its task of **path integration**. Many other insect navigators, spanning a spectrum from arctic bumble bees to temperate-zone honey bees to tropical orchid bees, are able to solve the same problem, but because they fly through cluttered environments rather than walk over open terrain, they are much less amenable to experimental study. This is not so for *Cataglyphis*, the desert ant. Careful experimental dissection of its navigational performances has shown that while foraging, the ant continuously measures angles steered

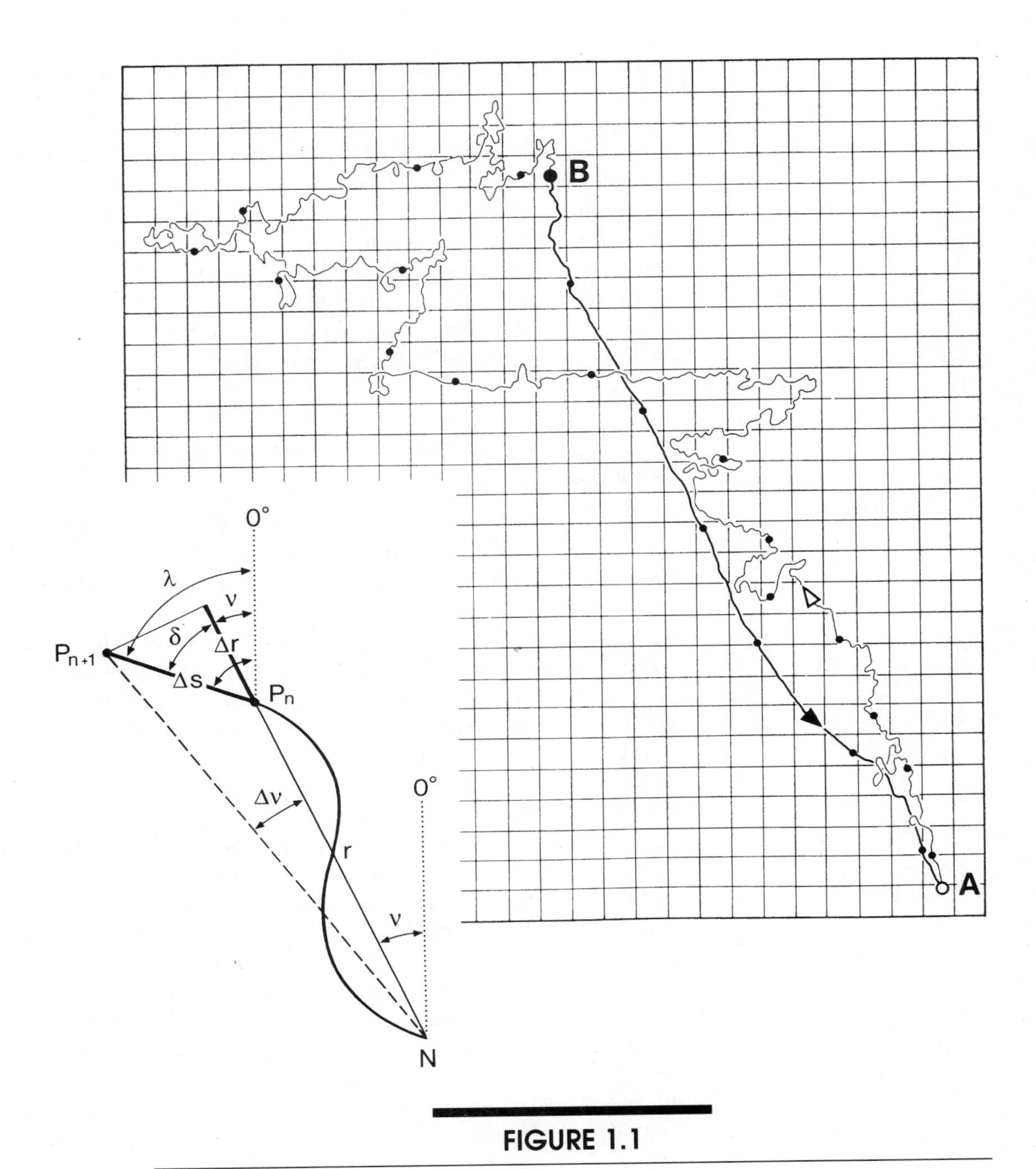

FIGURE 1.1

Outbound (foraging) and inbound (homing) path of a desert ant, *Cataglyphis fortis*, in a North African salt pan, the Chott-el-Djerid. Thin and thick lines mark the outbound and inbound path, respectively. **A**, nesting site; **B**, site of prey capture. The grid has a mesh width of 5 m. Small filled circles represent time marks noted every 60 s. The foraging run (path length 592.1 m) and the return run (path length 140.5 m) lasted 18.8 min and 6.5 min, respectively. Inset shows path integration. At P_n — the ant's actual position — the ant is informed about the home vector (r, ν). Locomotion in the direction λ by a path increment (Δs) results in the new position P_{n+1} characterized by the updated home vector (r + Δr, ν + Δν) as long as Δs is much greater than r. (For measurement of λ, see Figures 1.2–1.4.) (Adapted from Wehner and Wehner, 1990, and Hartmann and Wehner, 1995.)

and distances covered and integrates the angular and linear components of movement with remarkable acumen. Consequently, it always carries along in its mind a continually updated vector pointing to its home (see inset of Figure 1.1). This task is computationally demanding; *Cataglyphis* must solve the problem without straining its small nervous system. A human who moved about in a circuitous way on a featureless plain for distances of 10 to 20 km would encounter enormous difficulties to return directly, along a straight path, to the point of departure if not provided with a lot of instruments and some fairly detailed knowledge in mathematical calculus. The ant, however, has no access to any technical and computational tools (built, for example, into the inertial navigation system of modern aircraft) but nevertheless solves the problem amazingly well. How does this occur? Where is the compass, the odometer, the integrator?

Reading Skylight Patterns — A Simplified Model of the Outside World

The compass used by desert ants, honey bees, and probably many other insects is a skylight compass based on the **azimuthal** position of the Sun and, most effective, on a peculiar straylight pattern in the sky, the pattern of **polarized light** (called the **E-vector** pattern, as shown in both parts of Figure 1.2A). The scattering of sunlight by air molecules of the atmosphere causes the electric (E) vector of light coming from any particular point in the sky to oscillate in a particular direction (the E-vector direction). Because of intricate specializations in insect photoreceptors, the E-vector straylight pattern, invisible to humans, provides conspicuous visual cues to an insect navigator.

However, there are many E-vector patterns. During the course of the day, the Sun changes its position above the horizon and the E-vector pattern changes accordingly. (The only invariable feature is the symmetry plane formed by the solar **meridian**, that is, the perpendicular from the zenith through the Sun to the horizon, and its 180-degree counterpart, the antisolar meridian.) Nevertheless, within this changing celestial world, the insect can infer any particular compass course from any particular point in the sky! This task must be accomplished, for example, even under heavy cloud cover when the E-vector pattern can be read only in a small gap of clear sky. The prospect of solving this problem in a general, all-inclusive way is daunting. A physicist resorting to scientific first principles would have to run a sophisticated series of measurements and computations, take a number of linearly independent optical recordings in at least two pixels of sky, and would then have to use spherical geometry to perform elaborate three-dimensional constructions. Can we expect the insect's brain to do this? The answer, coming from a long and arduous series of experiments, is a clear no! Instead, insect navigators — the ant *Cataglyphis* and the bee *Apis*, for example — come programmed with a strikingly simple internal representation, a template of the external E-vector pattern. Although the actual pattern changes with the **elevation** of the Sun, the insect's inner replica of the pattern stays put.

How do we know this? We know because the insects consistently make mistakes if confronted with particular test situations. Such situations are created by letting *Cataglyphis* walk underneath a flat trolley that is moved along with the ant as it walks. The trolley serves as an optical laboratory by which the animal's visual surroundings can be manipulated experimentally. For example, the spectral and polarizational characteristics of skylight can be varied or the animal's view of the sky can be restricted to particular small aerial windows. If only a small window is available to view, ants trained to

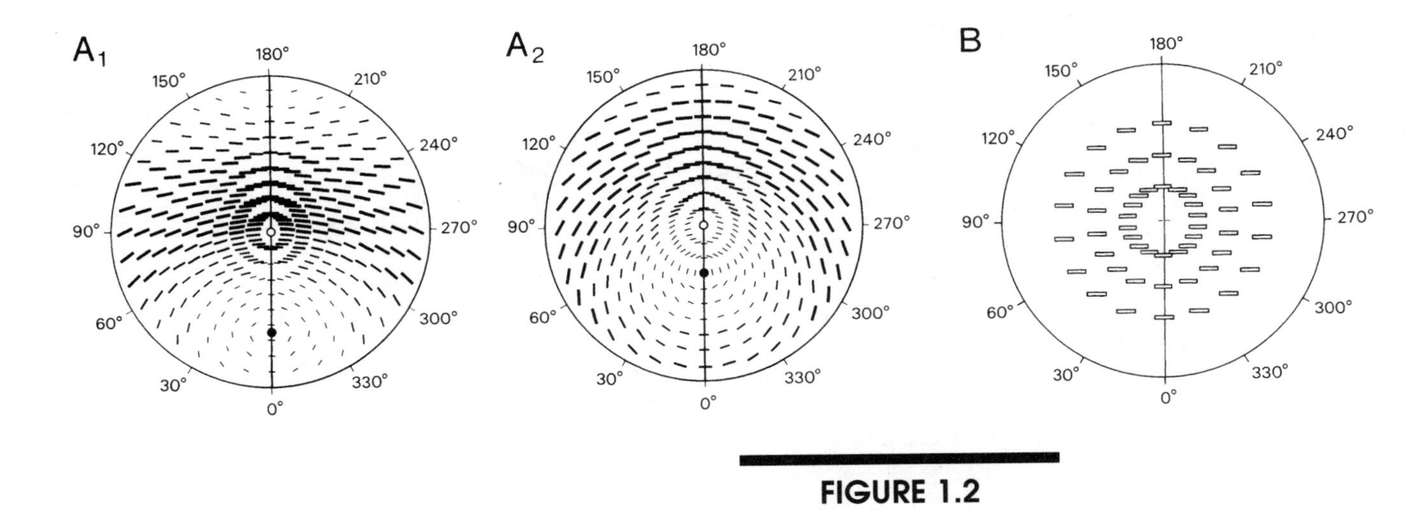

FIGURE 1.2

(**A**) Two-dimensional representation of the E-vector pattern in the sky shown for two elevations of the sun (black disc): 25° (A_1) and 60° (A_2) above the horizon. The E-vector directions are indicated by the orientation of the black bars. White disc marks the zenith. (**B**) The ant's stereotyped internal representation of the external E-vector patterns as derived from behavioral experiments. Points of the skylight compass are shown in the periphery. (Based on data from Rossel and Wehner, 1984, and Fent, 1986. For a summary of such experiments see Wehner, 1994.)

walk in a particular direction deviate by an error angle ε from this training direction, which means that the ants expect the particular E-vector direction to occur not at its actual position in the sky but to be shifted horizontally by (minus) ε relative to that position. Using this rationale, the insect's internal representation of the sky can be reconstructed point by point. The result is shown in Figure 1.2B. The stereotyped template resembles the skylight pattern when the Sun is at the horizon but differs from it for all other elevations. How can the insect navigate correctly by using an internal representation of the sky and not a correct copy of the external world?

Let us assume that the insect employs some kind of template-matching strategy. At any one time the best possible match between the internal template and the external pattern is achieved when the insect is aligned with the solar (or antisolar) meridian (Figure 1.3A). The distinction between the two principal meridians can be made by other means. The match (and its neural correlate) decreases as the animal deviates from the reference meridian, the zero point of its celestial compass. Using this strategy the animal should exhibit navigational errors whenever it is suddenly presented with an individual E-vector. Because of the discrepancy between the internal template and the external pattern, the individual E-vector is matched with the corresponding detector of the template only when the insect deviates by a certain angular amount from the solar meridian (Figure 1.3B). Consequently, navigational errors arise. Note, however, that such errors do not occur when the insect is continuously presented with the same patch of sky. The insect then uses the same reference direction of its compass, whether the reference point is the actual solar meridian or any other celestial meridian characterized by the current best match between the internal template and the outside world. In short, the mechanism is analogous to a magnetic compass that always has the needle erroneously pointed east rather than north — the compass is still as reliable as a normal compass.

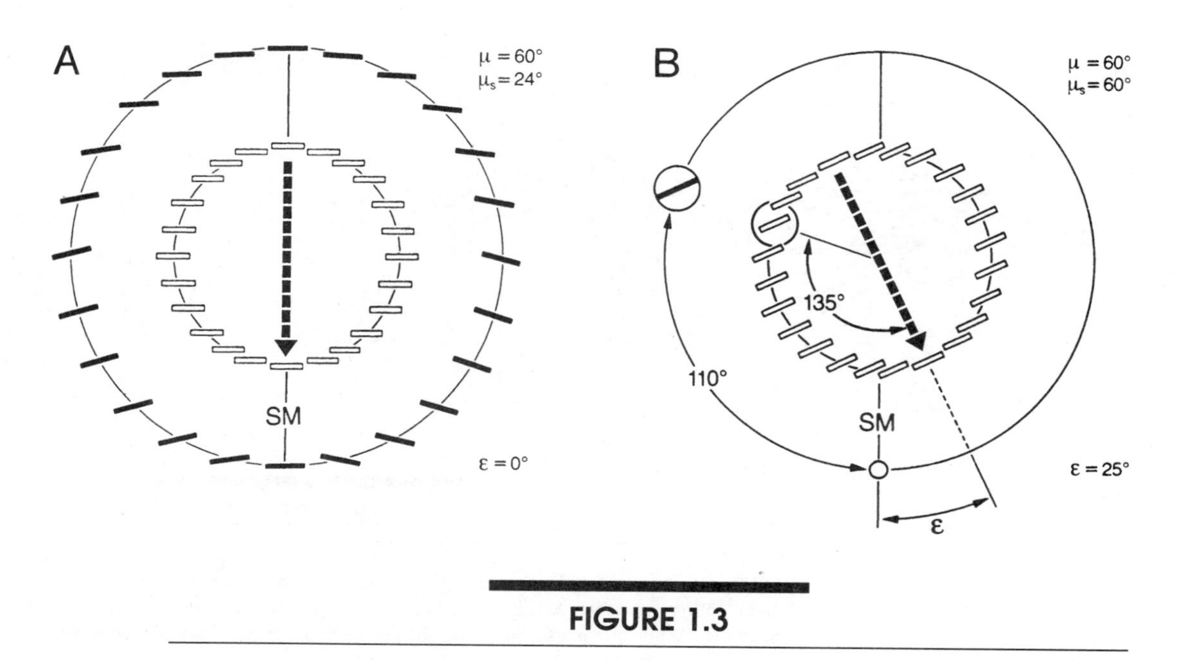

FIGURE 1.3

The ant's skylight compass: template-matching hypothesis. The array of E-vectors (in the external skylight pattern, indicated by filled bars) and the array of neural detectors (in the internal template, indicated by open bars) are shown on the outer and inner circles, respectively, for a particular parallel of altitude (μ) and a particular elevation of the sun (μ_s). The ants see (**A**) a full array of E-vectors or (**B**) an individual E-vector. Hatched heavy line symbolizes the insect's longitudinal body axis; ε is the error angle as observed in behavioral experiments; SM, the solar meridian. (Modified from Wehner, 1994.)

The characterization I have given here of the insect's skylight compass is avowedly sketchy. I have not referred to any aspect of the neural hardware used by the insect in processing E-vector information, nor have I outlined the different ways by which the insect might read its compass and how many points the compass might actually have. The point I would like to make is simply this: Evolution has managed to build into the insect navigator a nervous system that includes basic knowledge about the geometrical characteristics of the outside world, but this knowledge is sufficient if the navigator restricts its field trips to short periods of time. In fact, the insect assumes that the world does not change during any of its particular foraging excursions, that is, that the same patches of unobscured sky are present invariably. Given the ant's short foraging times (tens of minutes rather than hours), such an assumption is generally valid.

However, I have glossed over a major complication. The complication adds another layer of complexity to the problem. During the course of the day, the solar meridian (and with it, the E-vector pattern) rotates about the zenith and does so with nonuniform speed. This complication is serious because it implies that celestial cues do not provide the ant with the necessary geostable information. Foraging insects may return to the same feeding sites over periods of hours and days; thus, they must somehow be able to calibrate their compasses (for example, the position of the solar meridian) relative to an Earth-bound system of reference (Figure 1.4).

Recent evidence suggests that the landmark skyline at the horizon provides the necessary information. The calibration function — in astronomical terms, the **ephemeris function** — that correlates the azimuthal position of the solar meridian with the time of day is not constant; it varies with time of year and geographical latitude (Figure 1.5).

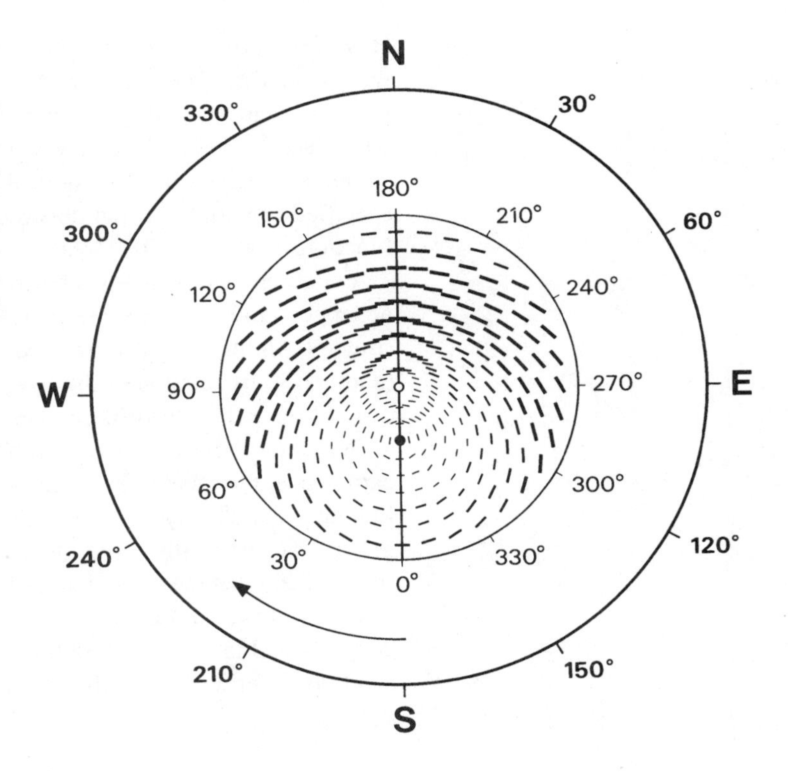

FIGURE 1.4

Calibration of the celestial compass (inner circle; cf. Figure 1.2A) against an Earth-bound system of reference (outer circle). The E-vector pattern rotates about the zenith (white disc) with daily westward movement of the Sun. Note that during rotation the Sun (black disc) changes its elevation above the horizon and, as a consequence, the E-vector pattern changes (cf. Figures 1.2A_1, A_2). The particular configuration shown here refers to local noon. The full calibration functions are given in Figure 1.5.

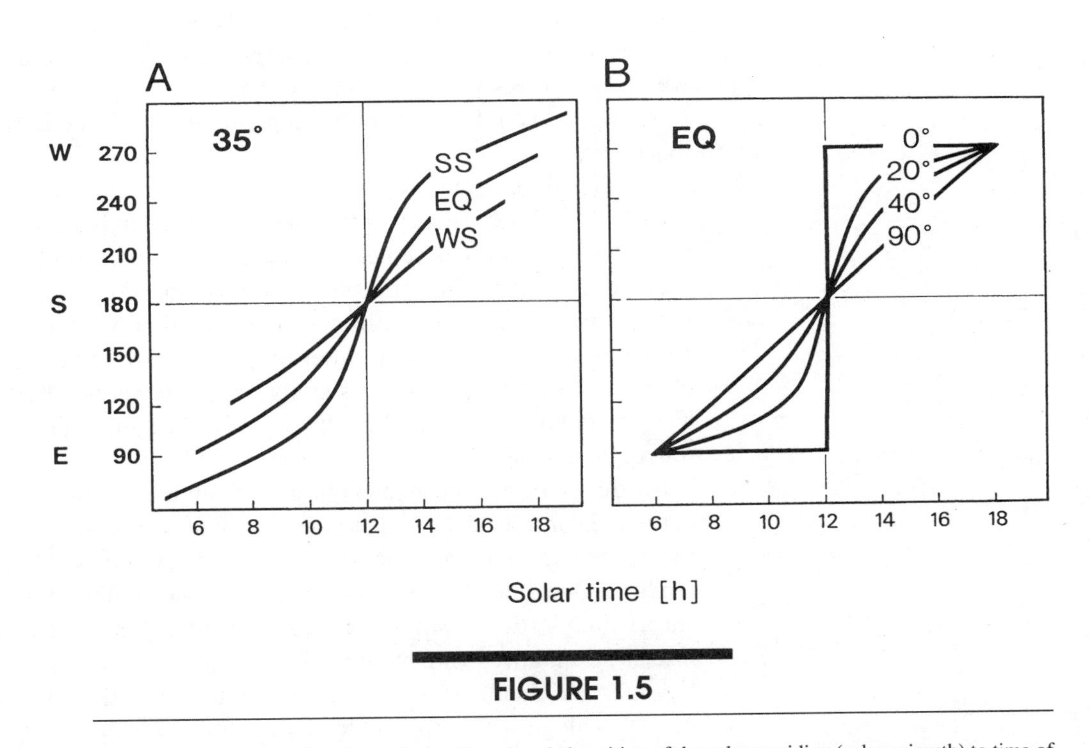

FIGURE 1.5

Calibration (ephemeris) functions relating the azimuthal position of the solar meridian (solar azimuth) to time of day (solar time). EQ, equinoxes; SS and WS, summer and winter solstices, respectively. (**A**) Geographical latitude (35°) constant, time of year varying. (**B**) Time of year constant (equinoxes), geographical latitude varying: 0°, equator; 90°, pole.

However, during their individual foraging lives, which last only days or at most a few weeks, ants and bees experience only a tiny fraction of the entire family of ephemeris functions and, thus, they encounter neither the possibility nor the need to acquire the whole set. It is equally unlikely that ants and bees are genetically programmed with something akin to an astronomical almanac in which all such ephemeris functions are specified or with the general astronomical and mathematical knowledge to derive the functions from first principles.

Experiments in which young foragers were tested at times of day at which they had never seen the sky before suggest that bees and ants use a simpler and more flexible strategy. All the insects seem to know inherently about the daily movement of the Sun (and hence the rotation of the E-vector pattern) is a basic rule: The azimuthal position of the Sun differs by 180 degrees between morning and afternoon. At local noon it switches abruptly from one position to the other. Later, based on an individual insect's experience, the innate 180-degree-step function is molded into a sigmoidal ephemeris function that closely resembles the one valid for the time and place concerned. Even though, at this writing, the mechanism by which this molding occurs is obscure, it seems likely that the insect interpolates at a uniform rate between individual fixes taken at various times of day. It would be worthwhile to study this phenomenon at higher geographical latitudes where the summertime ephemeris functions are rather smooth and thus differ most from the insect's innate 180-degree-step function.

Exploiting Landmark Panoramas — Maps in Insect Minds?

The sophisticated skylight compass is only one component of the path-integration system. Angles steered and distances covered must be integrated so as to provide the insect with a continuously updated home vector — an imaginary "safety line" without which *Cataglyphis* would be lost in the empty quarters of its desert habitat. But how safe is this line?

As all path-integration systems are open integrators, they are intrinsically extremely "noisy." While the insect integrates its path, navigational data are collected and processed within a self-centered (**egocentric**) rather than Earth-centered (**geocentric**) frame of reference. This process has unfavorable consequences: Errors accumulate as long as the animal moves. Errors can be erased only after the path-integration system has been reset to zero, that is, after the insect has returned to its starting position. Because of the cumulative error signals, safety decreases the farther the animal moves away from home. The tip of the home vector becomes blurred, so to speak, rather than remaining pointed.

At this point we must shift our gaze from the sky downward to the surface of the earth and consider the role landmarks may play in navigation. Even in desert habitats there are almost always shrubs, rocks, or stones to use as visual signposts to back up the noisy path-integration system — landmarks that are indeed used by insect navigators in surprisingly efficient ways. This efficiency has led some researchers to assume that insects are endowed with the remarkable ability to assemble map-like internal representations of the landmarks in their environment and then use such **cognitive maps** — mental analogues of topographic maps — to find their way to a familiar site even from points at which they have never been before.

However, as in Salome's dance, what lifting the veil suggests is more tempting than the view itself. Recent research has shown that ants and bees make intensive use of landmark information to relocate nesting and feeding sites and even routes between the two, but they do not incorporate such sites and routes into a map-like system of reference. Instead, the insects employ strategies that are more straightforward, foolproof, and largely sufficient for the task they must accomplish. If, for example, the task is to pinpoint the *Cataglyphis* nesting site after the path integration system has led the ant into close proximity of the goal and if the entrance of the nest is surrounded by an array of cylindrical landmarks and if the array is moved to a new territory, the ants displaced to the new area search with amazing precision and persistence for the now nonexisting nest. The mechanism works as if the ants had acquired an **eidetic image** — a "photographic snapshot" — of the landmark panorama around their nesting site and then move in the new area to match their individually acquired template as closely as possible with their current retinal image (Figure 1.6).

The matching-to-memory strategy can be studied best by distorting the training array and recording the insect's responses to its altered visual world. Even though in this experimental situation a full match between template and retinal image can no longer be achieved, there are always certain locations at which a better partial match occurs than at any other locations in the immediate neighborhood. The behavior can be modeled in computer simulations, in which the search behavior of the model insects resembles remarkably well the behavior of real ants or bees.

The snapshot concept seems to imply that the insect is rigidly bound to use landmarks only for recognizing one particular site. However, ants and bees can store panoramic images of numerous sites, take snapshots from various vantage points, and remember sequences of images that characterize entire routes of landmarks. They can even learn to follow different routes independently of vector information. Amazing as the landmark recollections may appear, it is evident that the insect navigators do not organize their experience on routes and sites into an integrated map of the landscape. Many lines of recent evidence suggest that bees and ants are not able to compute the spatial relationships among these routes and sites by mapping them within a common system of reference. This is neither a drawback nor altogether surprising. Insect foragers acquire landmark-based information in the context of path integration and thus are more or less restricted to the paths along which they have moved. They cannot take a bird's eye — or even bee's eye — view of the terrain over which they travel; instead, they assemble the necessary information piecemeal over time. The context-bound acquisition and retrieval of visual landmark information reduces the danger that the insect may become inappropriately trapped by similar landmark configurations present in the environment. Hence, it might be a safer and more robust strategy to rely on sequentially organized, gazetteer-like memories rather than to encode the spatial relations among a multitude of similar sites and routes in a large-scale mental map.

Experimental series started in the mid-1980s when the idea was first suggested that bees were able to form mental maps of their foraging ranges. What are the elementary building blocks of such maps? How could the elements be combined by a navigator that experiences them only successively and by egocentric perceptions? What might be the geometrical framework within which the spatial relations among these elements are represented?

The frame of reference used in the insect's fundamental mode of navigation — path integration (dead reckoning) — is egocentric. Like an invisible thread of Ariadne, the home vector (safety line) invariably ties the insect to the starting point of its foraging excursion, its central place, its home. Even though the safety line is continually reeled off as the insect returns home and thus its working memory is cleared of any vector

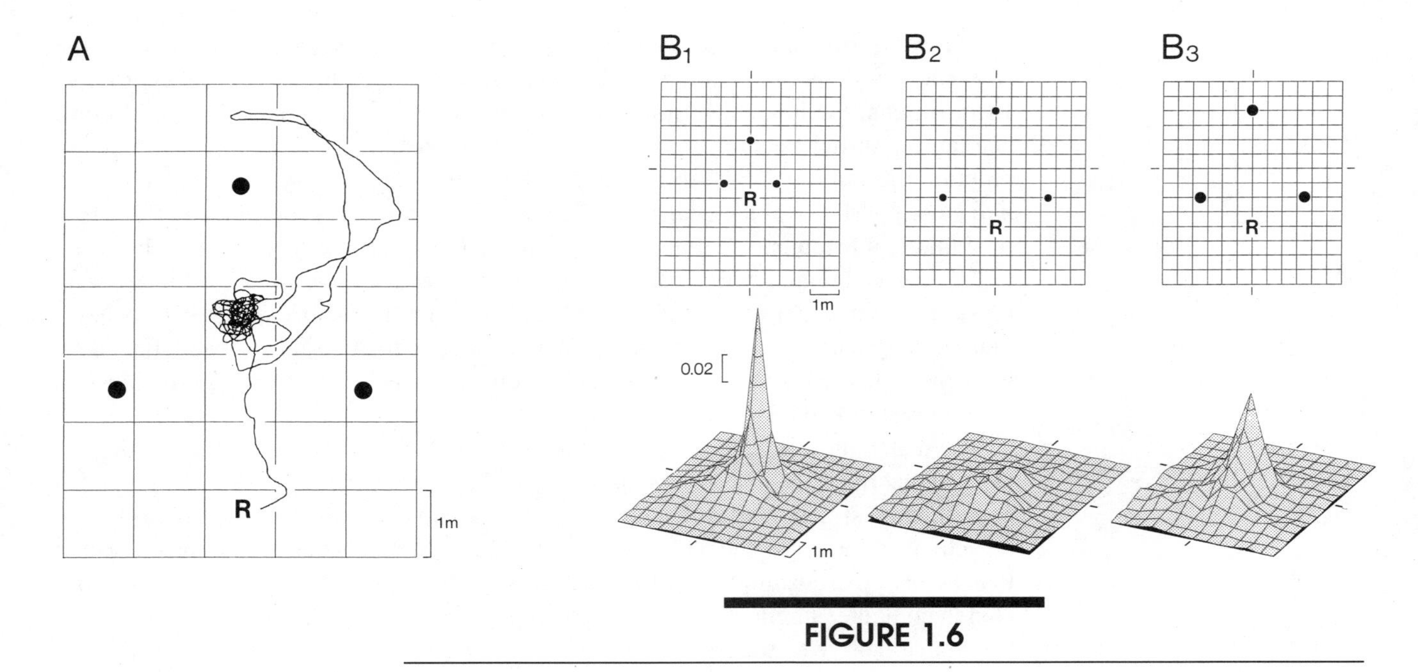

FIGURE 1.6

Navigation by landmarks (piloting) in the desert ant *Cataglyphis fortis*. (**A**) Search path of an individual ant trained to locate the center of a triangular array of cylindrical landmarks (11.5° high, 6.4° wide), shown by black discs. (**B**) Search density profiles (lower figures) of ants trained as in part A and tested under different stimulus configurations (upper figures): (**B$_1$**) Markers in same position as during training but displaced to new territory; (**B$_2$**) markers separated by twice the training distance; (**B$_3$**) markers twice the training size separated by twice the training distance; R, point of release. Diameters of markers in the upper figures of part B are increased by a factor of two. (From Wehner et al., 1996.)

information, the information persists in some higher-order reference memory. This conclusion can be drawn from the observation that ants and bees repeatedly return to previous feeding sites and that they do so even in featureless terrain by traveling along reversed home-vector courses (Figure 1.7A).

Furthermore, ants and bees acquire landmark-based information about homing and feeding sites as well as information about the routes traveled between such sites. They possess what could be called a landmark-based *route map* (Figure 1.7B). A landmark-based route is always coupled to a skymark-based vector, but the linkage can be broken experimentally. For example, if one rotates an artificial array of landmarks relative to the vector course, the ants are still able to follow the landmark route, even though the route is now oriented in a direction different from the one originally experienced. If, in addition, the insects were able to associate home-centered vectors with the landmark panoramas of various feeding sites, they could compute new routes by vector summation and subtraction and assemble a *vector map* (Figure 1.7C). But this they do not do, nor do they chart the positions of different sites within their foraging areas in a common *geocentric* frame of reference (for example, as a *metric map*, Figure 1.7D). Rather than computing the geometrical relations among familiar sites, they employ the flexible strategy of combining vector-, site-, and route-based information. Generally speaking, abstract floor-plan geometry gives way to a system of sequentially acquired and used matching-to-memory processes.

Systematic Search — The Last Resort

Suppose one clears all visual signposts from the ant's foraging ground — returning to the flat and featureless world of the supreme *Cataglyphis* navigator, the *fortis* ant. In this environment, the insect can no longer back up its path integration system by landmark guidance; it must resort to an emergency plan: A systematic search for the goal.

If lost, that is, if deprived of both vector and landmark information, *Cataglyphis* searches for home not by employing a random walk routine, but by switching on a more efficient **search program**. This program causes the ant to search in a system of loops of ever-increasing size, starting and ending at the origin. Consequently, the ant spends most of its time searching at locations where home is most likely to be, namely, at the tip of the home-based vector. What results is a markedly peaked search-density profile. This profile, however, is not rigidly fixed. Its width is matched to the size of the navigational errors inherent in the insect's path-integration system. The farther the ant has traveled, the larger are the errors that have accumulated during its foray. In accord with this prediction, the search-density profiles are wider and less peaked for long- rather than short-distance travelers. This is a nice adaptation of the insect's locomotor pattern to the expected spatial probability density function of the target — or, more generally, an adaptation of an insect's emergency plan to the conditions under which emergencies may eventually occur.

> God is a mathematician of a very high order, and He used very advanced mathematics in constructing the world.
>
> P. A. M. Dirac, *The Evolution of the Physicist's Picture of Nature*, 1963

LARGE-SCALE NAVIGATION — THE AVIAN MIGRANT

Dirac was wrong. The world, especially the living world, *evolved* rather than being constructed according to a mathematically advanced master plan; mathematical intelligence of a very high order is the *result* not the *cause* of the evolutionary process.

Vector Routes — Preordained Spatial Knowledge

A migratory bird covers much larger travel distances than does a foraging insect, but the difference is not merely one of scale. As we have seen, an insect acquires vector information while it performs its foraging trip. In contrast, a young bird that starts to migrate for the first time is already innately informed about its migration vector course. Recent research involving both large-range displacement experiments and tests of hand-raised birds in orientation cages has clearly shown that birds possess innate knowledge about their migration directions and distances, even about multileg **vector routes**.

One of the best studied cases is that of a European warbler, the blackcap, *Sylvia atricapilla*. Like many Palearctic long-distance migrants this warbler exhibits a fairly sharp "migratory divide." During fall migration, western German birds fly southwestward and eastern Austrian birds fly southeastward (Figure 1.8A and center of Figure 1.8B). The behavioral differentiation between the western and eastern European

FIGURE 1.7

Spatial representation by insect navigators. The ants and bees are central place foragers that routinely return to their central homing site (H). (**A**) Vector representation. Insects can learn vectors (vectors A_V, B_V) to at least two feeding sites (sites A, B). Vector information can be acquired and stored independently of local landmark information. (**B**) Landmark-based routes (A_R, B_R) are also learned. Sites H, A, and B are also characterized by memorized landmark panoramas (site-specific snapshots H_S, A_S, B_S). Grey patches symbolize landmarks. Spatial representation containing A_R and B_R could be called a route map.

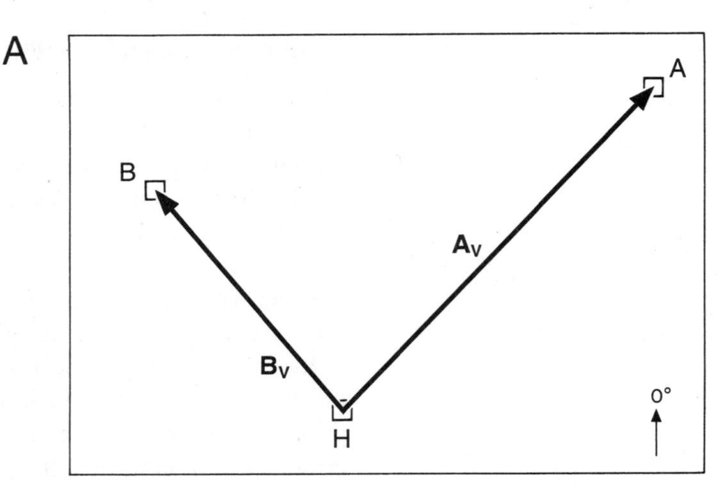

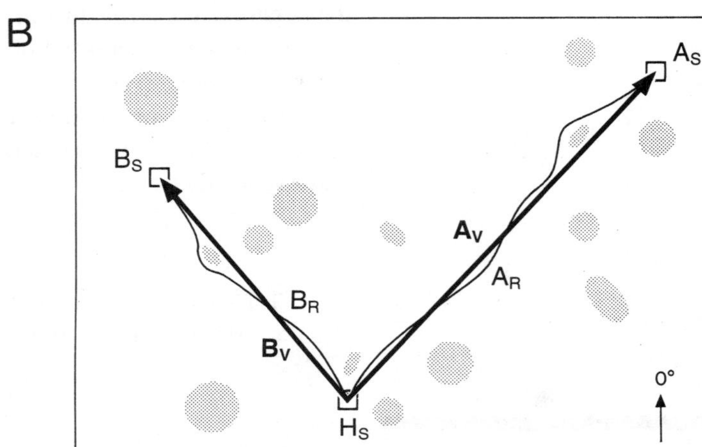

populations of blackcaps enables the birds to reduce the energy costs of migration by avoiding the topographical and ecological barriers of the Alps, the Mediterranean Sea, and the Sahara Desert. In fact, blackcaps are severely underrepresented among migrants crossing high Alpine passes.

The migratory divide of the two populations has been well documented by recoveries of ringed birds (to which small identifying ring-like leg bands have been attached), but, more important for the present context, the divide can be inferred directly from young, inexperienced, hand-raised birds that are tested in orientation cages at the time of day and year they would normally migrate. Within the cages, the birds hop or flutter in a direction that, on average, corresponds well with the migratory direction known from ringing experiments and exhibit the more migratory activity (as revealed by the number of hours and nights of "migratory restlessness") the longer their migration routes actually are. Such experiments have been repeated several times in a number of warbler species, and the results are now well documented (Figure 1.8B, circular diagrams). The experiments clearly prove that first-year migrants possess and use an innate vector program in setting their navigation courses, and this conclusion is further corroborated by crossbreeding experiments. The first-generation offspring of mixed pairs of southwestward and southeastward migrants choose migration directions that are intermediate between the routes of their parents (Figure 1.8C, D).

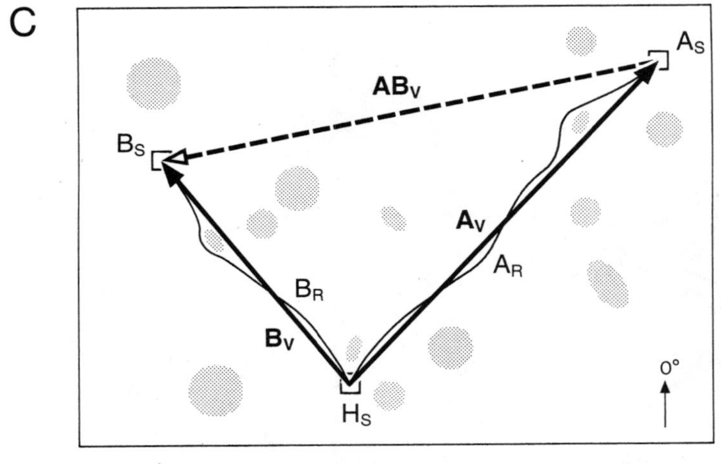

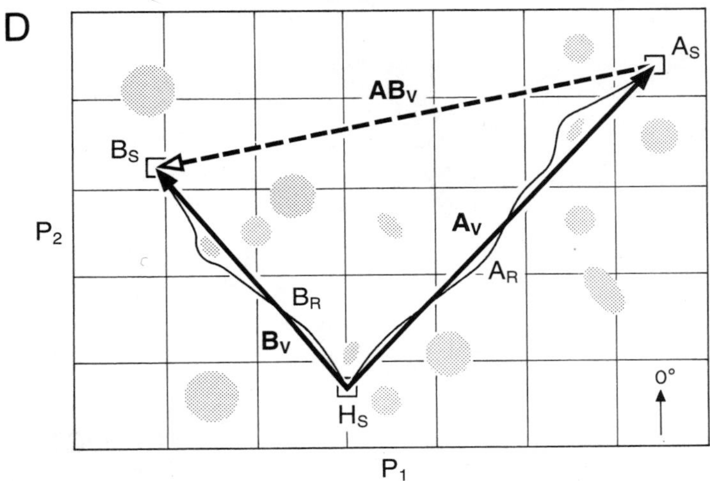

(**C**) Insects have not been shown to perform vector addition and subtraction to compute a new vector (AB_V) from two memorized vectors (A_V, B_V). Hence, the concept of a vector map is a hypothetical assumption. (**D**) A metric map concept in which vectors, sites, and routes are encoded in a general system of reference is also hypothetical. The reference is indicated by P_1 and P_2, but any other large-scale system of reference common to all landmark-based sites and routes could also apply. By use of such a common system of reference, the insect could compute new routes on the basis of landmark information. In summary, insects can assemble and use the spatial representations depicted in (A) and (B) but not the ones hypothesized in (C) and (D).

There is yet another twist in the story of the innate vector program. Hand-raised birds of the eastern population select southeastward directions during the months of September and October, but they shift to southerly directions when tested in November. The directional shift, exhibited by birds in home-based orientation cages, corresponds well with the natural migration route of the eastern population, which migrates first southeastward and then southward around the Mediterranean Sea to its winter quarters in East Africa. In contrast, the western population reaches its southwest European and northwest African winter quarters by following a fairly straight course and correspondingly does not show seasonal changes in its preferred directions when tested in orientation cages.

Orthodromes and Loxodromes — What Else?

Birds are able to use an inherited vector-orientation program. How do they put the program into action? On the basis of orientation-cage experiments, it is tempting to assume that the birds' migration program consists of a succession of vectors with directions and lengths defined, respectively, by compass courses and endogenous (internal) time signals. But what are the compasses and time signals? Are these navigational tools sufficient to solve the problem?

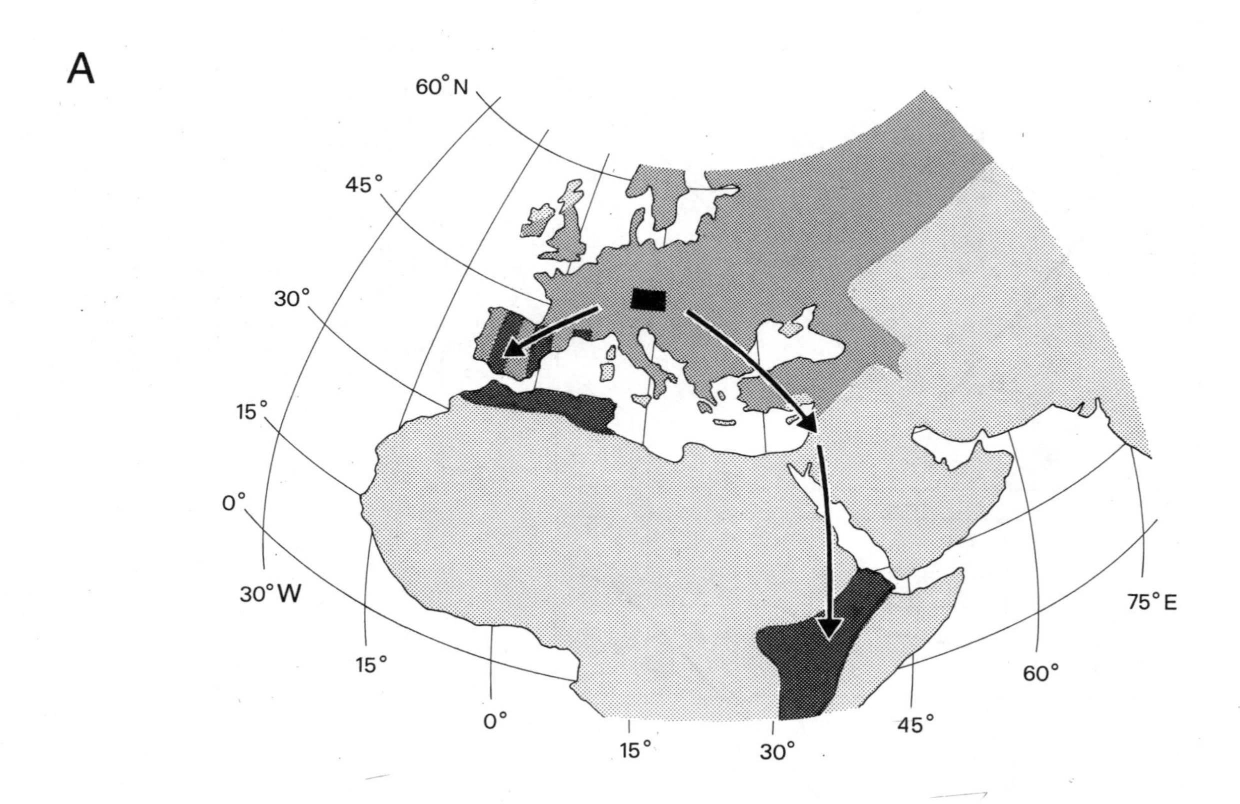
A
60° N
45°
30°
15°
0°
30° W
15°
0°
15°
30°
45°
60°
75° E

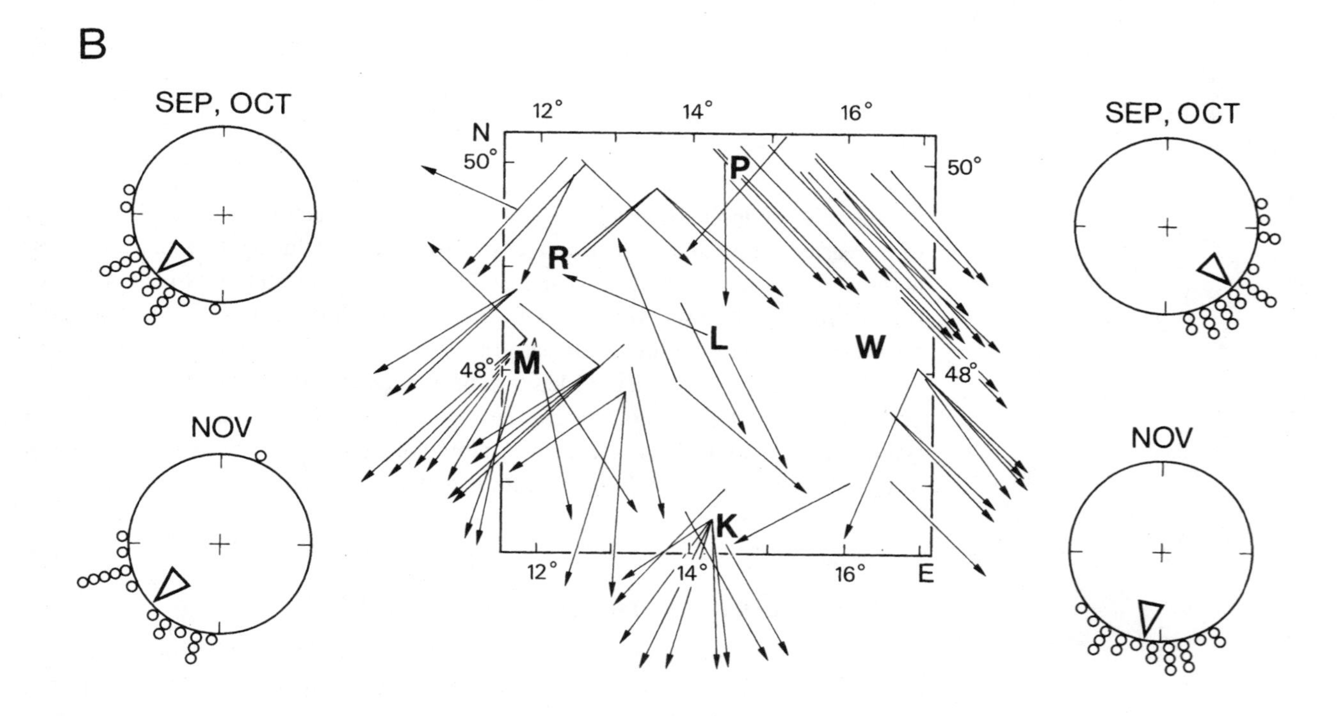
B
SEP, OCT
NOV
N
E
12°
14°
16°
50°
48°
P
R
L
W
M
K
SEP, OCT
NOV

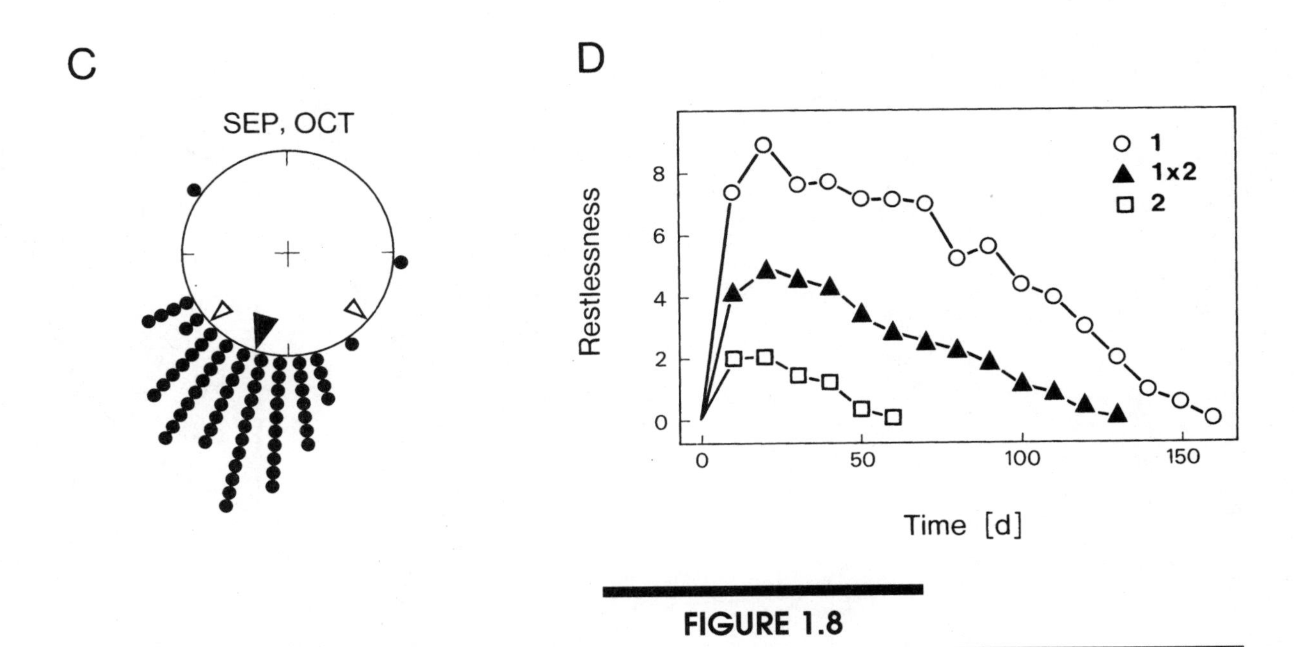

FIGURE 1.8

Migratory divide in European warblers, *Sylvia atricapilla.* (**A**) Map of breeding area (light grey), wintering area (dark grey), and major migration routes (arrows) as derived from ringing recoveries. Black rectangle marks area shown in middle of Part B. (**B**, *middle*) Recovery directions of birds ringed within an area of 400×400 km^2 (black rectangle in Figure 1.7A) and recovered more than 200 km away. K, Klagenfurt; L, Linz; M, Munich; P, Prague; R, Regensburg; W, Vienna. (**B**, *circular diagrams*) Directional choices of individual hand-raised birds taken from western (left diagrams) and eastern (right diagrams) populations and tested in orientation cages near Munich. Each data point (open circle) represents the mean direction chosen by an individual bird. Mean direction of all birds is indicated by open arrowhead. (**C**) Result of crossbreeding experiments showing that information about migratory directions is inherited. Filled symbols mark directional choices of first-generation offspring of pairs mixed from western and eastern populations bred in aviaries. Filled and open arrowheads indicate, respectively, the mean directions of the first-generation offspring and of the parental birds (for the latter, cf. part B). (**D**) Migratory restlessness displayed by birds taken from populations of southwest Germany (1) and the Cape Verde Islands (2) and tested in orientation cages near Munich. Data for first-generation offspring of mixed pairs from the two populations (1×2) are depicted by filled symbols. Birds of the Cape Verde Islands migrate due east to the North African wintering sites of the species. (Data from Berthold and Querner, 1981, and Helbig, 1991, 1994.)

To provide an answer we must take a closer look at the actual routes the birds follow during their migration from their summer breeding areas all the way to their winter survival sites, and vice versa. At first glance, this seems an impossible task, but, fortunately, modern technology provides tools suitable for recording the flight routes of migrating birds at different geographical scales, tools ranging from radar tracking in small areas to satellite-based radio telemetry for large-scale flight routes.

With all other things being equal (though they never truly are across the surface of our planet), the energetically least demanding way of traveling is to follow **great circle** (**orthodrome**) routes because these routes define the shortest distances between two points on the Earth's surface. In navigational terms, however, great-circle navigation is cumbersome. It implies that the migrants must continuously change their course as they intersect successive lines of longitude. On the other hand, following constant compass (*rhumbline* or **loxodrome**) courses is more convenient from an orientation point of view, but it leads to larger flight distances, especially at high geographical latitudes. In Figure 1.9A, orthodrome and loxodrome courses are shown on the surface of a globe. On two-dimensional maps, as they are usually used in human navigation, loxodromes

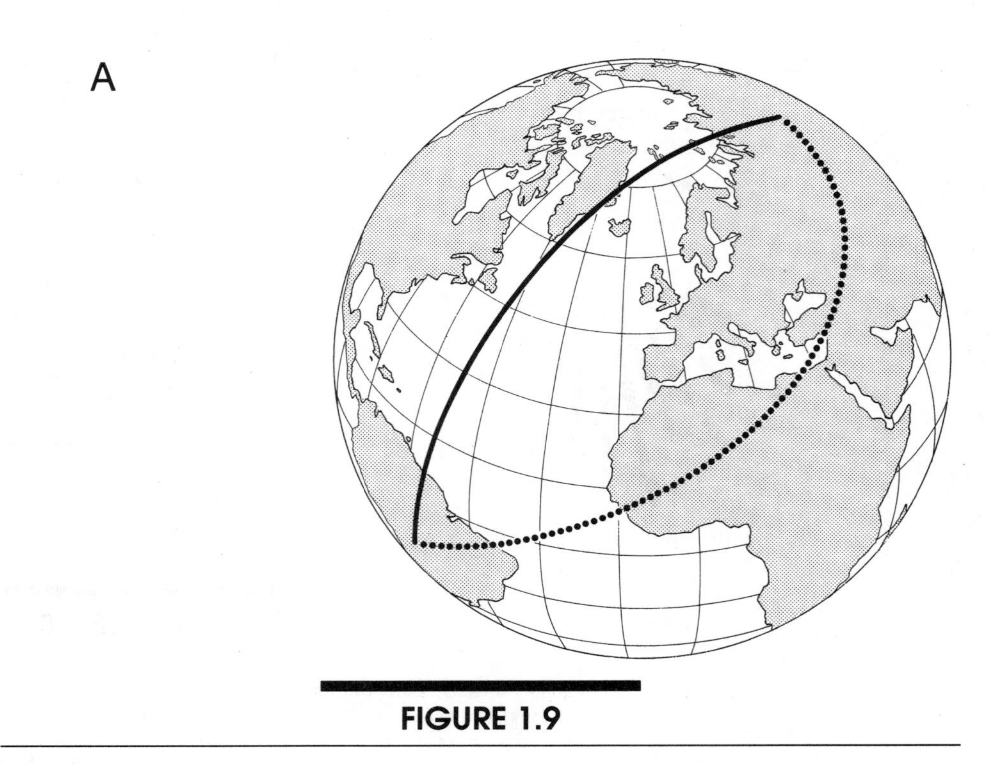

FIGURE 1.9

Orthodromes (solid lines) and loxodromes (dotted lines) shown in (**A**) three-dimensional and (**B**) two-dimensional representations. (B_1) Mercator (cylindrical) map projection; (B_2) azimuthal polar projection. Site A, Queen Elizabeth Islands; site B, Iceland; site C, Dutch Wadden Sea; site D, Taymyr Peninsula. Orthodrome (great-circle) distance from A to B is 2535 km; from C to D, 4234 km; loxodrome (rhumbline) distance from A to B is 2665 km; from C to D, 4634 km. (Map projections in part B based on Alerstam, 1990.)

are depicted as straight lines only if the map is based on a cylindrical projection (for example, the classic Mercator map) (Figure 1.9B_1). If, instead, great circles are meant to appear as straight lines, one needs a gnomonic projection (similar to the azimuthal polar projection in Figure 1.9B_2). Birds, however, travel without such maps or charts and without any idea in their minds that the Earth is a globe.

How, then, are birds able to perform? What do their actual flight patterns and migration routes look like? It has already become apparent from the cage experiments on European warblers that birds do not necessarily follow the shortest distance (great-circle) routes. A much more striking and, on first sight, counterintuitive, example is provided by North American warblers, the fall migration routes of which have been recorded by a network of radar stations. These warblers (in particular, the best studied blackpoll warbler, *Dendroica striata*), like many other passerine species and an abundance of waders, fly from their breeding areas in Canada and the northern United States south to the northern coast of South America. Surprisingly, however, on departure from the coasts of Nova Scotia, Cape Cod, and other parts of New England, they do not follow a southerly (great circle) route; they fly out over the Atlantic Ocean in a southeasterly direction. If they continued in that direction, they would end up in West Africa, not South America. But they do not. South of the Bermuda Islands they turn right, swinging on to a southwesterly course, and then, after having completed a 3,000- to 4,000-km, nonstop flight within 80 to 90 hours, they approach the Antilles and the coast of Venezuela from the northeast (Figure 1.10A).

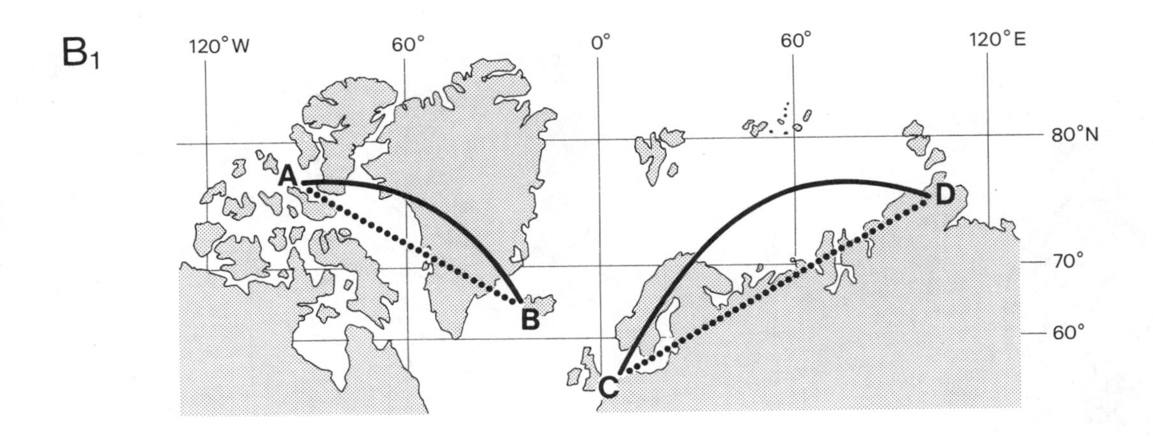

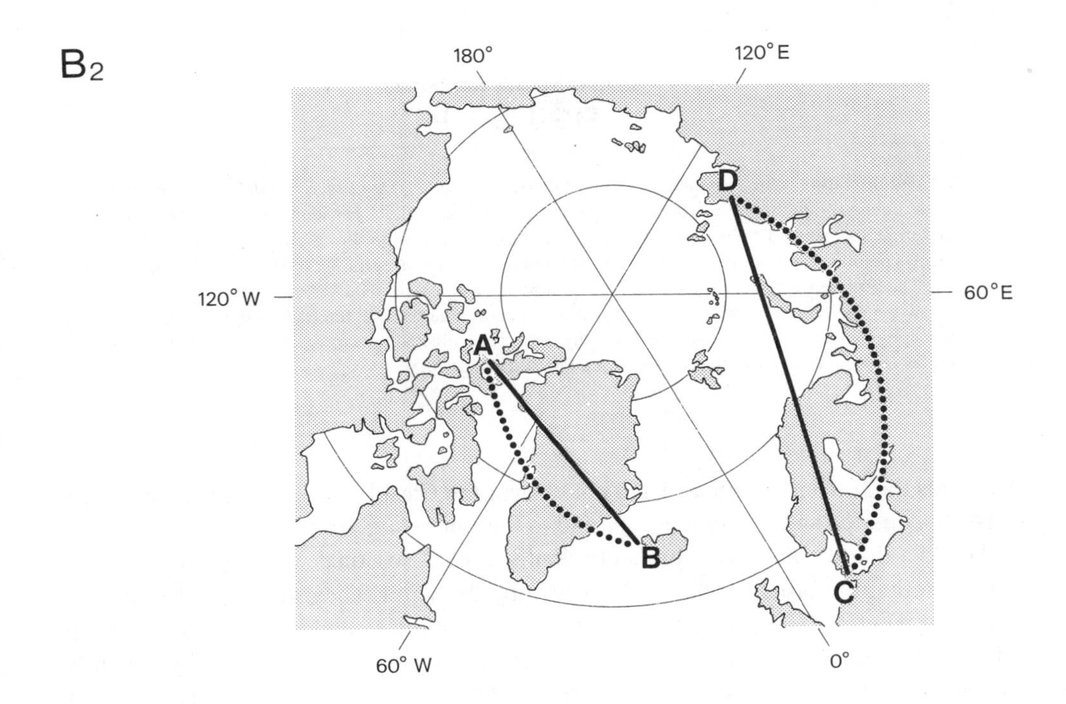

Perplexing as it might appear, the wide sweep — the vast detour across the Atlantic — is energetically the most economic route because this route is an adaptation to the wind and **barometric pressure** patterns prevailing over North America and the western Atlantic during the fall season. The birds, passerines and waders alike, depart from North America immediately after a barometric depression (cyclone) has passed eastward to benefit from the northwesterly winds behind the passing cyclone. South of the Bermudas, about 1000 km off the coast, the cyclone becomes stationary. The birds now pass the area of the cold front associated with the barometric depression and reach the trade-winds zone with its stable and clear weather. There, just as the old shipmasters did, they pick up the ideal (northeasterly) tailwinds for the second leg of their journey (Figure 1.10B). Additional evidence of the correctness of this interpretation is provided by the return route the migrants take during the spring season. At that time they avoid the unfavorable wind conditions over the Atlantic and follow a more westerly course

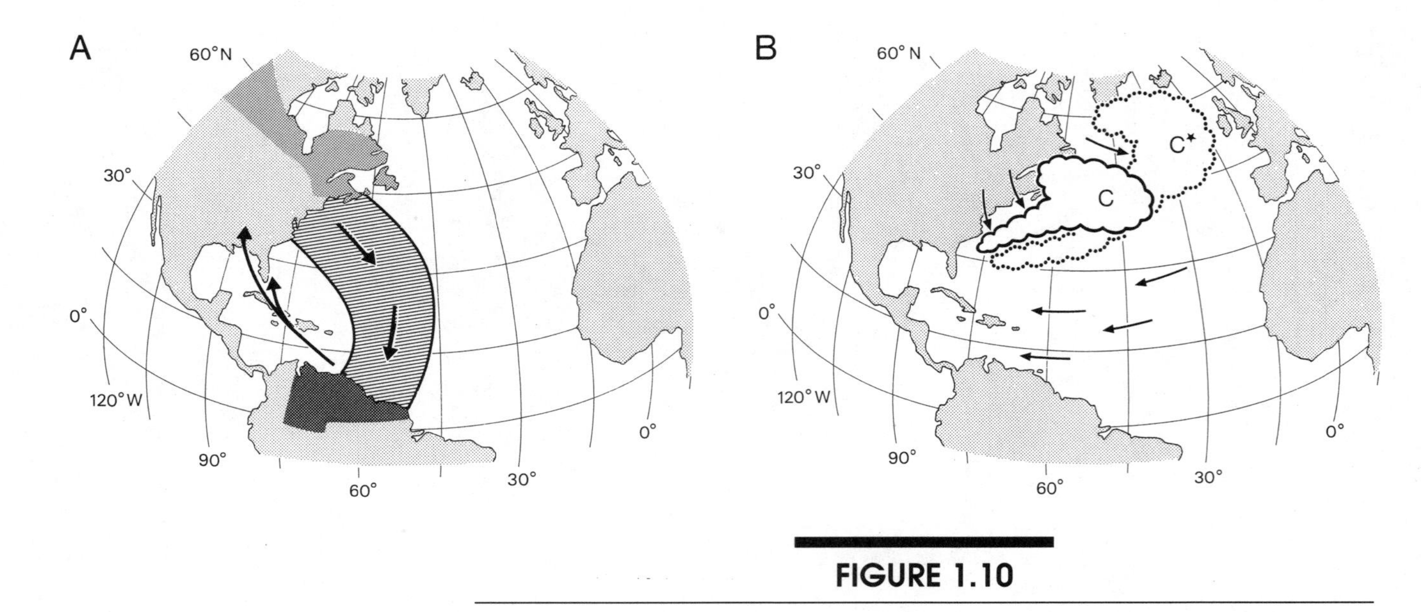

FIGURE 1.10

(**A**) Fall migration route (hatched) of North American warblers, *Dendroica striata*, from their breeding areas (light grey) to their wintering areas (dark grey). The same flight corridor is used also by many other passerine species and by waders. In spring, the birds return north along a more westerly flight path (arrows pointing northwestward). (**B**) Synoptic weather patterns over the West Atlantic during the fall season. The birds leave North America when a cyclone (C) has just passed over the New England coast; they follow the cyclone as it moves farther east (C*) and continue through the trade-winds zone, where they pick up favorable easterly tail winds. Wind directions are indicated by arrows. (Based on McNeil and Burton, 1977, Williams and Williams, 1978, and Richardson, 1979.)

that leads them from the Venezuelan coast, over the Caribbean Sea, to the south coast of North America and from there overland to their breeding grounds.

The fact that the migrating birds efficiently exploit seasonal weather patterns does not mean that the birds are passively driven by the wind. They must decide when to depart, at what altitude to fly, how to adjust their air speed, and what heading to keep relative to the direction of the wind. That these decisions are actually made by the birds can be inferred from extensive studies performed by Dutch researchers on many species of Siberian waders. During spring migration the plovers, knots, sanderlings, dunlins, godwits, and whimbrels — to name just a few of the species studied — travel along the **East Atlantic Flyway** from their West African wintering sites, the coastal mudflats of Mauritania, Guinea-Bissau, and Sierra Leone, to their arctic breeding areas (Figure 1.11). The birds fly at altitudes higher than 3 km. Only at that height do they encounter the favorable tailwind components that are necessary to keep the flight costs within reasonable ranges. Furthermore, they adjust their flight vectors (their headings and air speeds) relative to the wind vector in such a way that the resulting track vectors follow more or less the great circle route, which leads them to the next staging site, the Dutch Wadden Sea, 4300 km from the point of departure. They accomplish this task by heading out into the Atlantic (345° in Figure 1.12A, inner circle, that is, in the direction of Greenland). In other words, they adjust their flight vectors so as to allow themselves to be wind drifted in the proper track direction (19° in Figure 1.12A, outer circle), that is, toward the northwest European staging sites. If, instead, they headed directly for their

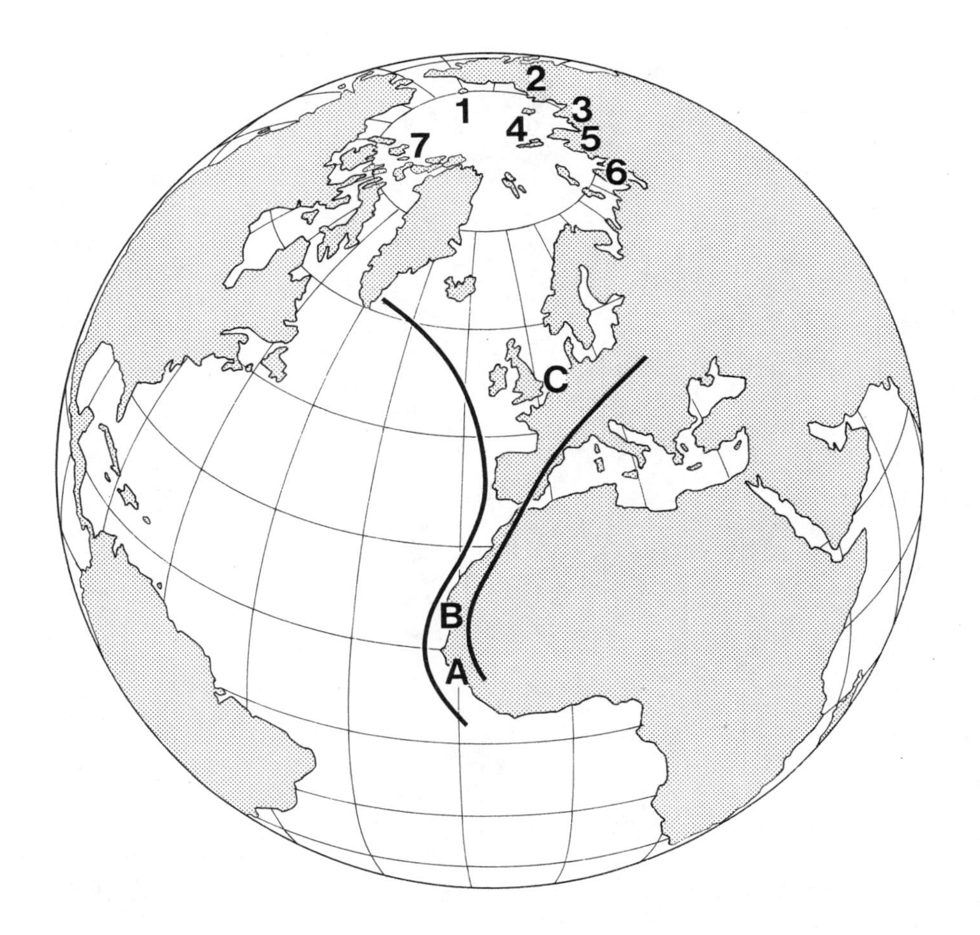

FIGURE 1.11

East Atlantic Flyway used by Siberian waders during spring migration. A, B, wintering sites; C, staging (stopover) site; 1–7, breeding sites. A, Guinea-Bissau; B, Banc d'Arguin (Mauritania); C, Wadden Sea (The Netherlands); 1, Wrangel Island; 2, Indigirka River; 3, Lena River; 4, New Siberian Islands; 5, Taymyr Peninsula; 6, Yamal; 7, Ellesmere Island. (Adapted from Wymenga et al., 1990.)

destination (if their flight angles and track angles coincided), they would be wind drifted at remarkable speeds straight eastward into the Sahara Desert.

We do not yet know what compass system — whether celestial or magnetic — the waders use in setting and maintaining their flight or track angles. Neither do we know whether they follow loxodrome or orthodrome courses. Even if we were able to record their routes by satellite tracking (and we are not, because with existing transmitters and batteries this recording technique is restricted to birds with body masses of more than 1 kg), the type of course would be difficult to decide because along the East Atlantic Flyway both courses are nearly identical. The courses differ markedly, however, at higher latitudes, for example, in polar regions, where the great circle distance may be as much as 36% shorter than the rhumbline distance.

It is in these polar regions that Thomas Alerstam and his associates, in a beautifully painstaking way, deciphered a very special record of how brent geese, *Branta bernicla*, reach their breeding grounds in northern Canada from spring stopover sites in Iceland. The geese do not take the direct (orthodrome) route, which would lead them over the steepest and widest parts of the Greenland ice cap, nor do they follow a constant (loxodrome) course. Instead, satellite tracking reveals a somewhat circuitous migration route — an adaptation to the large-scale topography of the area concerned. The geese leave Iceland in a west-northwesterly direction (with an average departure course of 295°); turn west-southwest when they reach the pack-ice zone off the steep coast of East

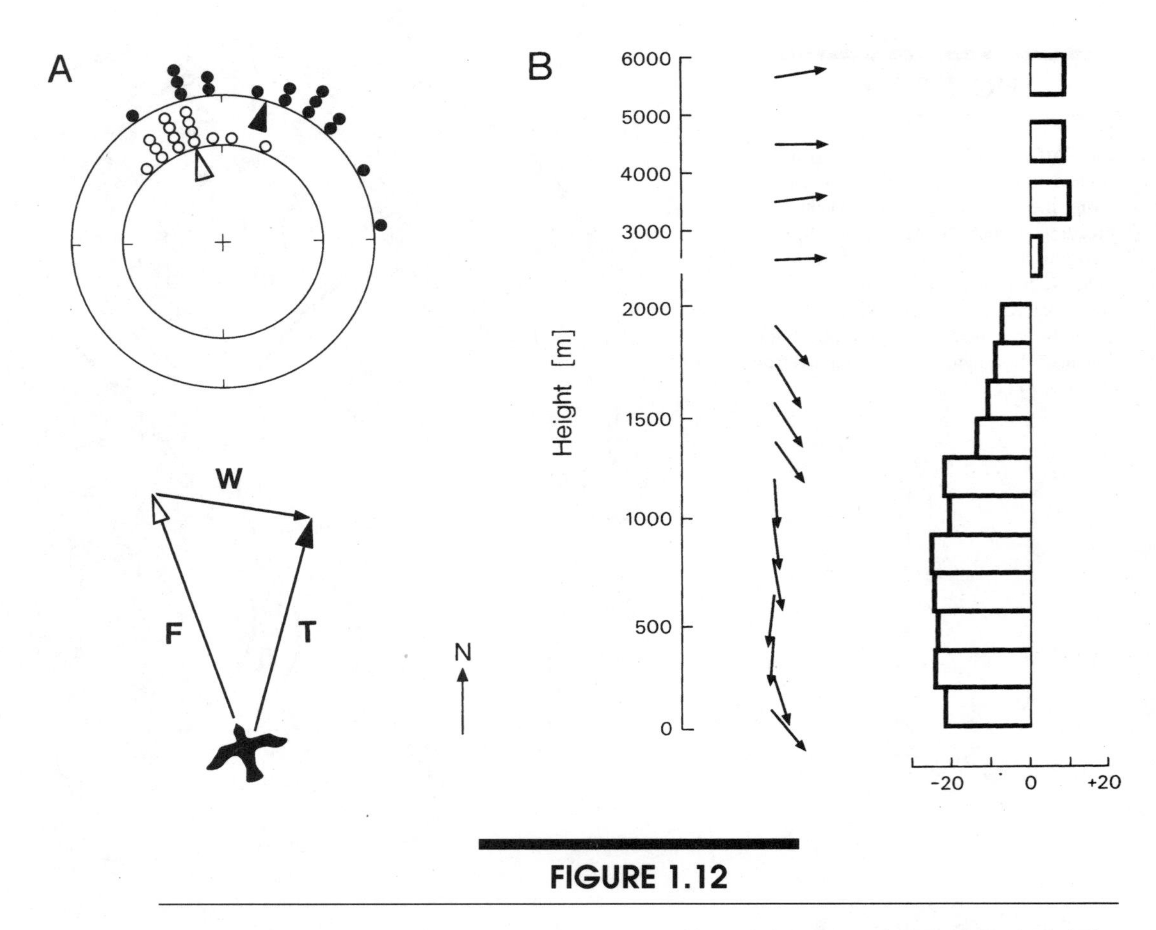

FIGURE 1.12

(**A**) Departure tracks and headings of waders leaving West African wintering sites (site B in Figure 1.10) during spring migration. (**A**, *upper*) Heading and track angles shown, respectively, on inner and outer circle. Individual circles are directions of individual flocks; total means are indicated by arrowheads; (**A**, *lower*) Scheme of the bird's air speed and heading direction (flight vector, F), wind speed and direction (wind vector, W), and ground speed and track direction (ground or track vector, T). (**B**) Average wind speeds and wind directions encountered by waders during ascent from the West African coast to altitudes of 6 km. (Data combined and modified from figures in Piersma et al., 1990.)

Greenland; stop for 2 to 7 days in a rather delimited area (at about 65°N, 38°W); and then continue across the Greenland ice cap at a course nearly identical to the one taken at Iceland (having a mean angle of 297°, Figure 1.13). It is a fascinating possibility that the geese use their temporary halt in East Greenland to reset their internal (**circadian**) clocks from local Icelandic to local Greenlandic time and then continue on the same Sun-compass course they have followed previously on their way to Greenland. We know from laboratory studies in other species of birds (starlings and homing pigeons) that under exposure to a new (time-shifted) 24-hour day/night regime it usually takes 3 to 6 days to recalibrate the Sun compass.

Carrying this exercise a bit further, let us imagine what would happen if a migrant bird did not stop every now and then to reset its internal clock, that is, to recalibrate its Sun compass when it is flying across different time zones (longitudes). Its flight bearings would then continuously change: The bird would intersect different lines of longitude at different course angles. This strategy might not be as unwise as it appears. Under

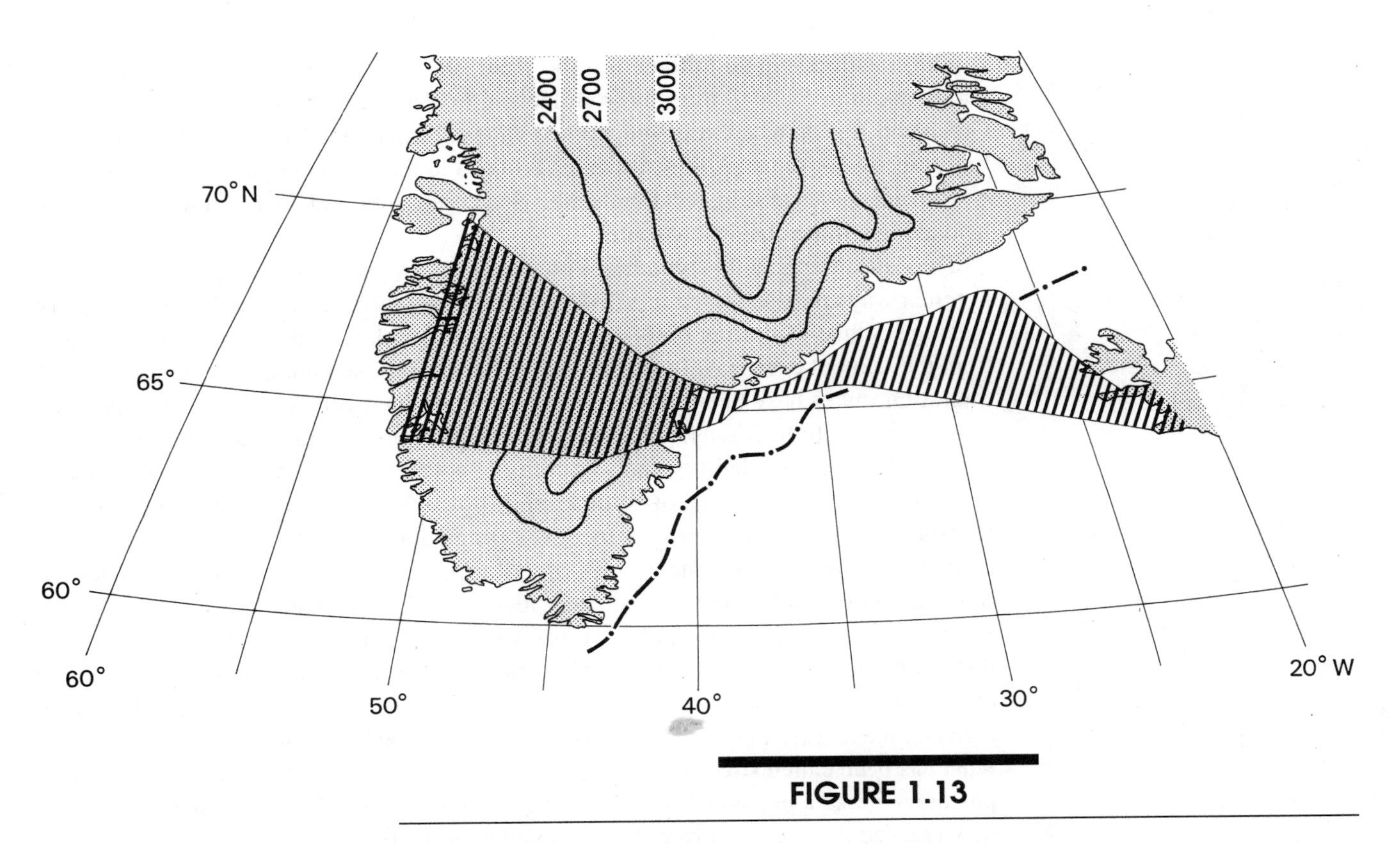

FIGURE 1.13

Flight corridor (hatched) used by brent geese, *Branta bernicla*, during migration from spring stopover sites in Iceland across Greenland toward breeding sites in northern arctic Canada. The dashed and dotted lines mark pack-ice limit. Contour lines within the Greenland ice cap are noted in meters. (Based on tracks of 10 geese registered via satellite-based radio telemetry by Gudmundsson et al., 1995.)

these conditions, the birds would travel quite precisely along a great-circle route. Even though this strategy of choosing a great-circle course might seem a figment of the human mind, it could actually be used by migrant birds in polar regions where the Sun remains above the horizon throughout the 24 hours of the polar summer day.

In this context, let us consider some recent radar studies of the migration of Siberian waders over the Arctic Ocean. At the end of the breeding season, the birds head due east over the vast expanse of pack ice. As radar trackings show, they take the great circle route to Alaska, their first stopover site on their global journey to their winter quarters in South America. If their directions recorded over the Arctic Ocean were extrapolated under the assumption that they kept a constant (loxodrome) course by continuously recalibrating their Sun compass or by relying on a magnetic compass system of one kind or another, they would end up in northernmost Canada, far distant from their fall migration route. Exposed to the harsh selection pressures of a global long-distance journey, the arctic waders might have learned, through evolutionary experience, that following a constant Sun-compass course but not continuously recalibrating the compass system while crossing different time zones would bring them via an energetically least-demanding route to their first postbreeding staging sites. Because in the Arctic there are no topographical or ecological barriers to cross, the great-circle route is indeed the one favored most in terms of energy expenditure.

The situation is quite different for the brent geese described earlier. If they followed the great-circle route all the way from Iceland to Canada, they would have to cross the Greenland ice cap where it is steepest and widest. Heavily loaded with fatty tissue (their fuel reserves for the trip), they are much reduced in their climbing capacities. For them, it is advantageous to stop, recalibrate their Sun compasses, and then start to cross Greenland along a more convenient route.

The data are in, but the hypotheses on the strategies that avian navigators use in finding their way are based more on advocacy rather than direct evidence. Nevertheless, the message is this: Migrating birds do not employ a grand, all-purpose system of navigation; they do not even travel for very long distances on fixed compass courses. Instead, migration routes have been shaped during evolution by a number of different **selection pressures,** including seasonal wind and barometric pressure patterns, large-scale topography, local availability and suitability of celestial or magnetic cues, and, though less understood, on the use of some kind of position control system. The birds have responded to selection pressures by developing sophisticated tools of migration and some truly intelligent means to integrate all the elements into a consistent system of orientation that enables them to end up with coherent travel plans.

In carrying out their travel plans, birds do not follow rigidly fixed routines. Instead, they continuously adapt their flight itineraries to environmental contingencies, employ emergency plans and backup systems, and gain new information about where they are at a given time relative to where they have been before. The latter capability leads finally to what has been called **site-based navigation**, which enables the birds to determine their position — take a positional fix — by using cues at a particular site irrespective of how they reached the site. For example, when young birds (first-year migrants) are displaced from their normal fall migration route by hundreds of kilometers and then released, they fly in their normal migration direction and arrive finally at places that lie completely outside the regular winter quarters of their breeding population. It is as if they had not realized that they had been displaced. Experienced birds (second-year and older migrants) behave differently. They alter their migration direction and return to their traditional wintering areas. In so doing they apparently have been able to place the point of release — a point on Earth at which they have never been before — on some kind of map and have worked out where they are relative to their known winter destination.

These conclusions are drawn from a classic series of experiments performed in starlings during a period longer than 10 years. However, if the ability to carry out site-based navigation is taken as a measure of performance, the accolade for success must be awarded to seabirds such as albatrosses, which in experiments have been displaced more than 5000 km from their breeding grounds but have been able to return to the exact sites within only 10 days. The geophysical cues (magnetic, infrasonic, or even olfactory) on which such maps might be based are hotly debated and are still elusive.

THE COLLECTIVE INTELLIGENCE OF THE NAVIGATOR'S MIND

What insects and birds accomplish in navigation are truly sophisticated tasks. If a human succeeded in designing computational programs that solved the underlying navigational problems, the person would certainly be regarded as highly intelligent. The fact that we hesitate to attribute the same degree of intelligence to ants and bees or to warblers and waders immediately indicates that intelligence cannot be inferred from the final product, from the result of the neural operations performed by the brain, but must be deduced from the computational structure of the operations themselves. What, then, is so particular and peculiar in the bees' and birds' procedures?

The peculiarity lies in the intricate interlocking of a great number of special-purpose **subroutines**. The insect forager and the avian migrant do not resort to first-principle mathematical and physical solutions when they negotiate their ways through cluttered local environments or over vast stretches of the surface of the Earth. They do not have any idea in their minds that the Earth is a globe, that there are geometrical and algebraic ways of computing great-circle courses across the surface of the globe or of reading E-vector directions from the celestial hemisphere.

Their subroutines used in navigation are intricate adaptations tailored to particular environmental conditions. Birds and insects seem to have dissected a complex, high-level task into a number of digestible, low-level bits. For example, the compass used in the insect's path-integration routine is based on a simplified model of the sky; the bewildering complexity of the insect's three-dimensional world of landmark configurations is decomposed into a temporal sequence of two-dimensional panoramic images. Of course, the strategies are based on simplifying assumptions about the spatial and temporal aspects of the navigator's visual surroundings, but the insect's behavior guarantees that the conditions to which the assumptions apply are usually met.

Because of their special-purpose status, the particular subroutines provide the navigator with only approximate or partial solutions to what the human investigator is inclined to envisage rationally as a rather general task. Nevertheless, in spite of the apparent limitations, the navigator's particular control systems are amazingly successful whenever the environment is regular and, hence, predictable enough to favor such evolutionarily preordained solutions. When environmental complexity and unpredictability rise for a given control system, other systems exploiting other sets of environmental parameters come into play either supportively or alternatively, in parallel or in succession.

In the "cockpit" of the navigator — its brain — different neural pathways are employed to mediate different control systems, and it is the collective interaction of the various **modules** that sets the stage for intelligent behavior to arise. Intelligence has not evolved as a unitary process or overriding factor that adds to the efficiency of the workings of rather specialized systems. Rather, intelligence is an emergent property of the cooperation among such systems.

Human beings are able to solve navigational problems of the sort described in this chapter by starting with first principles and then arriving at general-purpose solutions. On the basis of information gathered and transmitted nongenetically by members of our species for generations, humans have developed abstract concepts such as the concept of Newtonian space. Within this space humans feel free to perform whatever geometrical constructions they can think of and to develop the spatial conceptions used, for instance, in navigation. Bees and birds, however, find their way without embarking on such rational detours. Evolution has forced them to solve their problems in more immediate ways. But these immediate ways — the fundamental structures of spatial representation — that have evolved in insects, birds, and animals of all sorts, have not been lost in the genealogy of living beings. These structures remain in humans and underlie intuitions about the concepts we form in considering spatial relationships and solving spatial tasks. Metaphorically, the inner environment in which humans generate these concepts forms a prestructured landscape of considerable complexity rather than an even playing field; it is the lowly navigators highlighted in this chapter to which humans owe this complexity — this "strange intelligence."

FURTHER READING

General Surveys

Papi, F. (Ed.). 1992. *Animal Homing* (London: Chapman and Hall), 390 pp.

Wehner, R., Lehrer, M., and Harvey, W.R. (Eds.). 1996. *Navigation. J. Exp. Biol. 199*: 1–261.

Small-Scale Navigation: Insects

Cartwright, B.A. and Collett, T.S. 1983. Landmark learning in bees. Experiments and models. *J. Comp. Physiol. 151*: 521–543.

Cartwright, B.A. and Collett, T.S. 1987. Landmark maps for honeybees. *Biol. Cybern. 57*: 85–93.

Collett, T.S. 1992. Landmark learning and guidance in insects. *Phil. Trans. R. Soc. Lond. B 337*: 295–303.

Dyer, F.C. 1991. Bees acquire route-based memories but not cognitive maps in a familiar landscape. *Anim. Behav. 41*: 239–246.

Dyer, F.C., Berry, N.A., and Richard, A.S. 1993. Honey bee spatial memory: Use of route-based memories after displacement. *Anim. Behav. 45*: 1028–1030.

Dyer, F.C. and Dickinson, J.A. 1994. Development of sun compensation by honeybees: How partially experienced bees estimate the sun's course. *Proc. Natl. Acad. Sci. USA 91*: 4471–4474.

Fent, K. 1986. Polarized skylight orientation in the desert ant *Cataglyphis*. *J. Comp. Physiol. A 158*: 145–150.

Gould, J.L. 1986. The locale map of honey bees: Do insects have cognitive maps? *Science 232*: 861–863.

Hartmann, G. and Wehner, R. 1995. The ant's path integration system: A neural architecture. *Biol. Cybern. 73*: 483–497.

Menzel, R., Chittka, L., Eichmüller, S., Geiger, K., Peitsch, D., and Knoll, P. 1990. Dominance of celestial cues over landmarks disproves map-like orientation in honey bees. *Z. Naturforsch. 45*: 723–726.

Ronacher, B. and Wehner, R. 1995. Desert ants *Cataglyphis fortis* use self-induced optic flow to measure distances travelled. *J. Comp. Physiol. A 177*: 21–27.

Rossel, S. and Wehner, R. 1984. Celestial orientation in bees: The use of spectral cues. *J. Comp. Physiol. A 155*: 605–613.

Srinivasan, M.V., Zhang, S.W., Lehrer, M., and Collett, T.S. 1996. Honeybee navigation en route to the goal: Visual flight control and odometry. *J. Exp. Biol. 199*: 237–244.

Wehner, R. 1994. The polarization-vision project: Championing organismic biology. In: Schildberger, K. and Elsner, N. (Eds.), *Neural Basis of Behavioural Adaptations* (New York: G. Fischer), pp. 103–143.

Wehner, R., Bleuler, S., Nievergelt, C., and Shah, D. 1990. Bees navigate by using vectors and routes rather than maps. *Naturwiss. 77*: 479–482.

Wehner, R. and Menzel, R. 1990. Do insects have cognitive maps? *Ann. Rev. Neurosci. 13*: 403–414.

Wehner, R., Michel, B., and Antonsen, P. 1996. Visual navigation in insects: Coupling of egocentric and geocentric information. *J. Exp. Biol. 199*: 129–140.

Wehner, R. and Srinivasan, M.V. 1981. Searching behaviour of desert ants, genus *Cataglyphis* (Formicidae, Hymenoptera). *J. Comp. Physiol. A 142*: 315–338.

Wehner, R. and Wehner, S. 1990. Insect navigation: Use of maps or Ariadne's thread? *Ethol. Ecol. Evol. 2*: 27–48.

Zeil, J. 1993. Orientation flights of solitary wasps (*Cerceris;* Sphecidae, Hymenoptera). II. Similarities between orientation and return flights and the use of motion parallax. *J. Comp. Physiol. A 172*: 207–222.

Large-Scale Navigation: Birds

Able, K.P. 1980. Mechanisms of orientation, navigation, and homing. In: Gauthreaux, S.A. (Ed.), *Animal Migration, Orientation, and Navigation* (New York: Academic Press), pp. 283–373.

Alerstam, T. 1990. *Bird Migration* (Cambridge: Cambridge Univ. Press), 420 pp.

Alerstam, T. 1990. Ecological courses and consequences of bird migration. *Experientia 46*: 405–415.

Alerstam, T. 1996. The geographical scale factor in orientation of migrating birds. *J. Exp. Biol. 199*: 9–19.

Alerstam, T. and Pettersson, S.-G. 1991. Orientation along great circles by migrating birds using a sun compass. *J. Theor. Biol. 132*: 191–202.

Berthold, P. 1973. Relationships between migratory restlessness and migration distance in six *Sylvia* species. *Ibis 115*: 594–599.

Berthold, P. (Ed.). 1991. *Orientation in Birds* (Basel: Birkhäuser Verlag), 331 pp.

Berthold, P., Helbig, A.J., Mohr, G., and Querner, U. 1992. Rapid microevolution of migratory behaviour in a wild bird species. *Nature 360*: 668–669.

Berthold, P. and Querner, U. 1981. Genetic basis of migratory behavior in European warblers. *Science 212*: 77–79.

Emlen, S.T. 1975. Migration: Orientation and navigation. In: Farner, D.S. and King, J.R. (Eds.), *Avian Biology*, Vol. 5 (New York: Academic Press), pp. 129–219.

Ens, B.J., Piersma, T., Wolff, W.J., and Zwarts, L. (Eds.). 1990. Homeward bound: Problems waders face when migrating from the Banc d'Arguin, Mauritania, to their northern breeding grounds in spring. *Ardea 78*: 1–36.

Gudmundsson, G.A., Benvenuti, S., Alerstam, T., Papi, F., Lilliendahl, K., and Akesson, S. 1995. Examining the limits of flight and orientation performance: Satellite tracking of brent geese migrating across the Greenland ice-cap. *Proc. R. Soc. Lond. B 261*: 73–79.

Gwinner, E. 1996. Circadian and circannual programmes in avian migration. *J. Exp. Biol. 199*: 39–48.

Helbig, A.J. 1991. Inheritance of migratory direction in a bird species: a cross-breeding experiment with SE- and SW-migrating blackcaps (*Sylvia atricapilla*). *Behav. Ecol. Sociobiol. 28*: 9–12.

Helbig, A.J. 1991. SE- and SW-migrating blackcap (*Sylvia atricapilla*) populations in central Europe: Orientation of birds in the contact zone. *J. Evol. Biol. 4*: 657–670.

Helbig, A.J. 1994. Genetic basis of evolutionary change of migratory directions in a European passerine migrant, *Sylvia atricapilla. Ostrich 65*: 151–159.

McNeil, R. and Burton, J. 1977. Southbound migration of shorebirds from the Gulf of St. Lawrence. *Wilson Bull. 89*: 167–171.

Perdeck, A.C. 1958. Two types of orientation in migrating starlings, *Sturnus vulgaris*, and chaffinches, *Fringilla coelebs*, as revealed by displacement experiments. *Ardea 46*: 1–37.

Piersma, T. 1994. Close to the edge: Energetic bottlenecks and the evolution of migratory pathways in knots. Ph.D. Thesis (Univ. Groningen), 366 pp.

Piersma, T., Zwarts, L., and Bruggemann, J.H. 1990. Behavioural aspects of the departure of waders before long-distance flights: flocking, vocalizations, flight paths and diurnal timing. *Ardea 78*: 157–184.

Richardson, W.J. 1979. Southeastward shorebird migration over Nova Scotia and New Brunswick in autumn: A radar study. *Can. J. Zool. 57*: 107–124.

Williams, T.C. and Williams, J.M. 1978. Orientation of transatlantic migrants. In: Schmidt-Koenig, K. and Keeton, W.T. (Eds.), *Animal Migration, Navigation, and Homing* (Berlin: Springer), pp. 239–251.

Wymenga, E., Engelmoer, M., Smit, C.J., and van Spanje, T.M. 1990. Geographical breeding origin and migration of waders wintering in West Africa. *Ardea 78*: 83–112.

CHAPTER 2

COMMUNICATION AND THE MINDS OF MONKEYS

Robert M. Seyfarth* and Dorothy L. Cheney*

WHAT IS SPECIAL ABOUT THE HUMAN MIND?

For centuries, humans have speculated about the differences between their own minds and the minds of nonhuman creatures, particularly monkeys and apes. For the most part, this speculation has focused on one attribute more than any other: language. To the ancient Greeks, who recognized that nonhuman primates were, anatomically, more like humans than were any other creatures, the crucial difference lay in the fact that man possessed **syntax** (Sorabji, 1993). Monkeys and apes (the Greeks made no distinction between them) were obviously clever, crafty, and exhibited almost every characteristic associated with human intelligence including the use of words, but they had no system of rules or grammar for assembling the words into meaningful phrases or sentences.

The Greeks' conclusion that language, and particularly syntax, was the defining attribute of *Homo sapiens* persisted throughout the Middle Ages and well into the eighteenth and nineteenth centuries. It was not, however, based on any scientific study of nonhuman primate behavior or communication; instead, the uniqueness of human language was simply taken for granted. In place of careful, scientific studies there arose, particularly in the nineteenth century, a wealth of popular, satirical accounts in which an ape is introduced into civilized human society and his surprisingly human behavior, but lack of human language, is used by the author to make fun of public figures. In Thomas Love Peacock's novel *Melincourt* (1817), for example, a young "orangutan from Africa" (he must mean a chimpanzee) is regarded as far superior to his human companions in native gallantry and nobility of feeling, at least in part, because he rescues a maiden in distress without taking advantage of her. Admittedly, the ape lacks

*Departments of Psychology and Biology, University of Pennsylvania, Philadelphia, PA 19104 USA

language and the facility of speech, but this hardly hinders his success as a Member of Parliament, where his silence "gives him the reputation of a powerful but cautious thinker" (Henken, 1940; Janson, 1952).

In the twentieth century, scientists have continued to focus almost exclusively on language as the defining feature of human minds and behavior. Their approach has, however, been somewhat more scientific. As a first step, language is broken into its constituent parts and each is examined separately. Once this is done, it becomes clear that there are at least two features that appear to distinguish language from other forms of animal communication. The first is **semantics**, defined as the use of signs or sounds to represent objects and events in the external world; the second is syntax (Jackendoff, 1994; Pinker, 1994).

To test the hypothesis that syntax, semantics, or some other property is in fact the crucial feature that distinguishes language and the human mind from the minds of monkeys and apes, scientists have embarked on two sorts of research. In one set of studies, an ape is brought into the laboratory and attempts are made to teach it a form of human language. The goal of this work is to see whether any nonhuman creature has the potential to learn language and, if so, to determine whether it can learn to communicate linguistically with the same ease as a human child (see Chapter 3). In a second set of studies, scientists went into the field, immersed themselves in the natural behavior and communication of free-ranging monkeys or apes, and tried to understand whether the animals' vocalizations, studied on their own terms, satisfy the defining features of human language. Field studies are the focus of the present chapter.

SYNTAX AND SEMANTICS

The natural vocalizations of monkeys and apes have now been studied for more than 20 years, and on the basis of the research there is no reason to believe that the Greeks were wrong about syntax. Although we still know surprisingly little about the vocalizations of our closest nonhuman primate relatives, the chimpanzee and the gorilla, no primate species' communication studied thus far provides any example of the kind of syntax found in human language (Robinson, 1984). Research on nonhuman primate semantics, however, paints a more complicated picture.

In East Africa, vervet monkeys (*Cercopithecus aethiops*) use a variety of acoustically different alarm calls when they encounter different predators (Struhsaker, 1967; Seyfarth et al., 1980a, 1980b). Each alarm call type elicits a different, apparently adaptive, response from monkeys nearby. For example, alarm calls given to leopards (*Panthera pardus*) cause vervets to run into trees, but alarm calls given to martial and crowned eagles (*Polemaetus bellicosus* and *Stephanoaetus coronatus*, respectively), sound quite different and cause vervets to look up or run into bushes. Alarm calls to snakes are also acoustically distinct and cause nearby animals to stand on their hind legs and peer into the grass around them. Finally, **playback experiments** (in which tape-recorded calls are played back to the monkeys) demonstrate that an alarm call alone, even in the absence of an actual predator, elicits the same responses as does the predator itself (Seyfarth et al., 1980b). In this respect, calls function in much the same way as human words do (Hockett, 1960; Seyfarth and Cheney, 1992). When one vervet hears another give an eagle alarm call, the listener responds just as if the animal had actually seen the eagle.

One is tempted to conclude that in the monkey's mind the call stands for or conjures up images of an avian predator even when the bird itself has not been seen. But is this really true? When does a sound become a word?

There are good reasons for caution in drawing parallels between vervet alarm calls and human words. Granted, the calls certainly seem to function like words in the monkeys' daily lives, but such appearances could easily be misleading. After all, for Pavlov's dogs the sound of a bell may have stood for or conjured up images of meat, but this does not prove that the dogs understood the relation between bells and meat in the same way that humans understand the relation between the word chair and a particular piece of furniture.

When does a monkey's call cease to become a sound and become a word? This happens, the psychologist David Premack suggests, when the properties ascribed to the call are not those of a sound but those of the object it denotes (Premack, 1976). Human language offers some excellent examples. When asked to compare the words treachery and deceit, humans ignore the fact that they sound different (that is, they have different acoustic properties) and describe them as similar because they have similar meanings. By contrast, treachery and lechery, despite their acoustic similarity, are treated as different because they mean different things. In making these judgments, humans recognize the referential relation between words and the things for which they stand. When comparing words, humans judge them to be similar or different not just on the basis of their physical properties but also on the basis of their meanings. Do the calls of vervet monkeys qualify as words in this stronger sense? Can we really claim that monkeys understand the meaning of their vocalizations?

To investigate how vervets compare vocalizations we borrowed a method from research on speech perception in human infants, called the **habituation/dishabituation technique**. The method is based on the observation that subjects — particularly human infants — who hear the same sound repeatedly gradually cease responding to the sound, that is, they habituate. However, if subjects who have habituated to one stimulus hear another that they judge to be different, the strength of their response increases sharply. The habituation/dishabituation technique offers a useful method for getting nonverbal organisms, like babies or nonhuman creatures, to reveal whether they judge two stimuli to be similar or different.

To test whether vervet monkeys compare vocalizations on the basis of their acoustic properties or their apparent meaning, we chose as stimuli two calls the monkeys give during territorial encounters with neighboring groups. The first, a long loud trill called *wrr*, is given when another group has first been spotted. It seems to alert the members of both groups that a neighboring group has been seen. The second vocalization, a harsh raspy sound called *chutter*, is given when an intergroup encounter has escalated into aggressive threats, chases, or fighting. *Wrrs* and *chutters* thus have broadly similar referents — they both seem to mean another group has been sighted — but are very different acoustically. As a result, vervets asked to compare *wrrs* and *chutters* in a habituation/dishabituation experiment should judge them to be different if the comparison were based on acoustic properties but similar if the comparison were based on meaning.

To begin our experiments we first selected a subject. Then, on day 1, as we followed the animals around their group's range, we played the subject a particular adult female's *chutter* and filmed the subject's response. This control test allowed us to establish the baseline strength of the subject's response to the vocalization. Then, on day 2, the subject heard the same adult female's *wrr* repeated eight times at roughly 20-minute intervals. Because no other group was present, we predicted that the subject would rapidly habituate to the call. Finally, roughly 20 minutes after the last *wrr*, the subject heard the female's *chutter* again (the test condition). If subjects judged *wrrs* and *chutters* to be similar calls (that is, if the monkeys compared calls on the basis of meaning), habituation to an individual's *wrr* should also produce habituation to the same individual's *chutter* and there would be a large decrement in response strength between control and test conditions. On the other hand, if subjects judged *wrrs* and *chutters* to be different (that is, if the monkeys compared calls on the basis of acoustic properties and not meaning), habituation to an individual's *wrr* should not produce habituation to the same individual's *chutter* and there would be little difference in response strength between control and test conditions.

Results provided clear evidence that vervet monkeys compare different calls on the basis of meaning, not just by acoustic properties. When subjects were presented with the same individual's *wrrs* and *chutters*, they transferred habituation across the two call types. In other words, if a subject had habituated to individual X's *wrr*, the subject also ceased responding to individual X's *chutter*. By contrast, when subjects were played two calls whose referents were different, they did not transfer habituation across call types. If a subject had ceased responding to individual Y's leopard alarm call, the subject still responded at normal strength to individual Y's eagle alarms (Cheney and Seyfarth, 1988).

Compared with our earlier experiments on the vervets' alarm calls, these tests addressed the question of meaning and reference more directly by asking animals to compare two vocalizations (that is, to judge them to be either similar or different) and to reveal the criteria they use in making their comparison. The results suggest that when one female vervet monkey hears another vocalize, she forms a representation, in her mind, of what that call means. And if, shortly thereafter, she hears a second vocalization, the two calls are compared not just according to their acoustic properties but according to their meanings. If we accept the notion that a monkey's call becomes a word when the properties ascribed to the call are not those of a sound but those of the object it denotes, *wrrs* and *chutters* seem to become words.

Vervet monkeys — and probably many other nonhuman primates — thus present us with a rudimentary semantic system in which some calls, such as leopard and snake alarms, are markedly different in meaning, but others, such as *wrrs* and *chutters*, are linked to a common referent and can be used to represent shades of meaning within a general class. Drawing on our knowledge of the vervets' ecology and social behavior, we can also suggest why monkeys need such communication and hence why the cognitive abilities that underlie it — in this case, the ability to treat sounds as signs that represent things — may have evolved. If different predators demand different escape strategies, the adaptive value of acoustically different alarm calls is obvious. In addition, monkeys often vocalize when out of sight of each other, which favors the evolution of calls whose meaning can be derived from acoustic features alone and does not depend crucially on contextual cues, such as what the listener sees. Further, the appropriate response to a vocalization may differ markedly from one individual to the next. A monkey on the ground who gives an eagle alarm may be looking up into the sky, but others nearby should run into bushes and those in a tree should run down and out of the tree.

These circumstances favor callers who can communicate about events in ways that are relatively independent of their own behavior, and listeners who can interpret a call's meaning in a manner that is relatively independent of what the caller is likely to do next. They favor, in other words, semantic, representational communication.

THE ATTRIBUTION OF MENTAL STATES

But do the results really elevate monkey calls to the status of human words? Human semantics, after all, involves more than just an understanding of the referential relation between words and the objects or events they denote. When communicating, humans also attribute mental states such as knowledge, beliefs, or desires to others and recognize that mental states have causal power: What an individual thinks influences what the person does.

To attribute beliefs, knowledge, and emotions to others is to have what psychologists call a **Theory of Mind** (Premack and Woodruff, 1978). This concept is regarded as theory because mental states are not always directly observable; thus, if one person attributes a mental state to another, the person cannot do so on the basis of what is actually perceived but must, instead, act on the basis of a theory about what the other person is like.

In a classic demonstration of the development of a theory of mind in children, the psychologists Hans Wimmer and Josef Perner (1983) presented 3- to 9-year-old subjects with scenarios in which the children had to describe the knowledge of others. In one case, the children watched a puppet show in which a boy, Maxi, puts a piece of chocolate into a blue cupboard. Maxi then leaves the room, and in his absence his mother removes the chocolate from the blue cupboard and places it in a green one. The children were asked where Maxi would look for the chocolate. Children younger than 4 years of age consistently answered "the green cupboard" — the cupboard in which they themselves knew that the chocolate was located. In contrast, about half the 4- to 6-year-old children and more than 80% of the 6- to 9-year-old children correctly pointed out that Maxi would still think that the chocolate was in the blue cupboard. The younger children's errors were not due to a failure of memory, because most of the children who gave an incorrect answer to the question nevertheless gave a correct answer when asked if they remembered where Maxi had put the chocolate. Instead, it seems as if the children's ability to attribute knowledge to another and to recognize that another person's knowledge might be different from their own (and indeed incorrect) does not become established until roughly the ages of 4 to 6 years.

Can animals distinguish between their own beliefs and the knowledge and beliefs of others? Do animals, for example, ever take special steps to inform an individual who is ignorant or to correct another's false belief? To take a closer look at these questions and to get a better idea of why the attribution of mental states represents a crucial watershed in the evolution of language and human intelligence, consider some recent research on the function of vocalizations in baboons.

Grunts That Appease and Reconcile

Only a small proportion of the vocalizations given by monkeys and apes occur in the form of alarm or intergroup calls. Instead, the most common calls given by many nonhuman primates are quiet grunts, coos, or trills that are uttered at close range and occur in the context of social interactions or group movement. Many calls appear to function to initiate and facilitate social interactions. For example, in Japanese macaques (*Macaca fuscata*), grooming interactions are often initiated when one female vocalizes to a potential partner (Masataka, 1989; Sakuro, 1989). In stump-tailed macaques (*Macaca arctoides*), individuals that grunt to mothers before attempting to handle their infants are less likely to receive aggression than are individuals that remain silent (Bauers, 1993).

From a functional perspective, such calls are interesting because animals grunting or cooing to one another look so much like humans engaged in conversation. Typically, there is no obvious response to the calls from nearby listeners, and it certainly seems as if these vocalizations, like many human conversations, function simply to mediate social interactions and "grease the social wheels." Notice, however, that it is difficult for humans even to describe how monkey calls function without assuming that the subjects have a theory of mind. If a call serves to mollify an opponent or a subordinate mother, it seems — to humans — almost essential that the caller be able to recognize the partner's anxiety and to signal its own benign intent. But is this really the case?

To examine the function of close-range vocalizations in more detail, we recently carried out a detailed study of the grunts given by free-ranging female baboons (*Papio cynocephalus ursinus*) in the Okavango Delta of Botswana. Our work focused on a group that included about 70 individuals: adult males, adult females, and immature baboons. We were particularly interested in the grunts that were exchanged between the 23 adult females (Cheney et al., 1995).

Like adult females in many species of Old World Monkeys, female baboons form stable linear dominance hierarchies (Seyfarth, 1976; Hausfater et al., 1982; Smuts and Nicolson, 1989). Although most affinitive interactions occur among close kin, adult females also interact with unrelated females, particularly if the females have infants. Normally, if a dominant female approaches a subordinate female, the subordinate is supplanted and moves away. Frequently, however, the dominant female vocalizes to the subordinate by using a low-pitched, tonal grunt. The grunts seem to have an appeasing function because they increase the probability of a subsequent friendly interaction, such as grooming or infant handling.

We recorded 2698 incidents in which one female approached another that ranked lower than herself; in 621 (23%) of the situations, the dominant female grunted to the subordinate. There were 17 females that could approach at least one lower-ranking unrelated individual. For 15 of the 17, the mean frequency of approaches to all possible partners that was followed by a friendly interaction was higher if the dominant female first grunted than if she did not (Figure 2.1A). Similarly, for 14 of 17 individuals, the mean frequency with which a female supplanted her lower-ranking partner was higher when she did not call than when she did (Figure 2.1B). Results were unaffected by the relative difference in rank between the two females. Grunts, therefore, appear to mediate and facilitate social interactions among unrelated adult females.

If grunts or other vocalizations do function to facilitate affinitive interactions, they might also be expected to play a role in reconciling opponents following aggression. Nonhuman primates are frequently aggressive toward one another, yet they live in relatively stable, cohesive social groups. Recent studies have suggested that opponents may mollify the effects of aggressive competition by reconciling soon after fighting or threatening one another (de Waal and van Roosmalen, 1979; de Waal and Yoshihara, 1983; York and Rowell, 1988; Aureli et al., 1989; Cheney and Seyfarth, 1989; Judge, 1991; Aureli, 1992; Cords, 1992, 1993). Two animals are said to have reconciled if, within minutes of behaving aggressively, they interact in a friendly way by touching, hugging, grooming, or approaching one another. No study, however, has yet considered the role that vocalizations might play in reconciling former opponents.

FIGURE 2.1

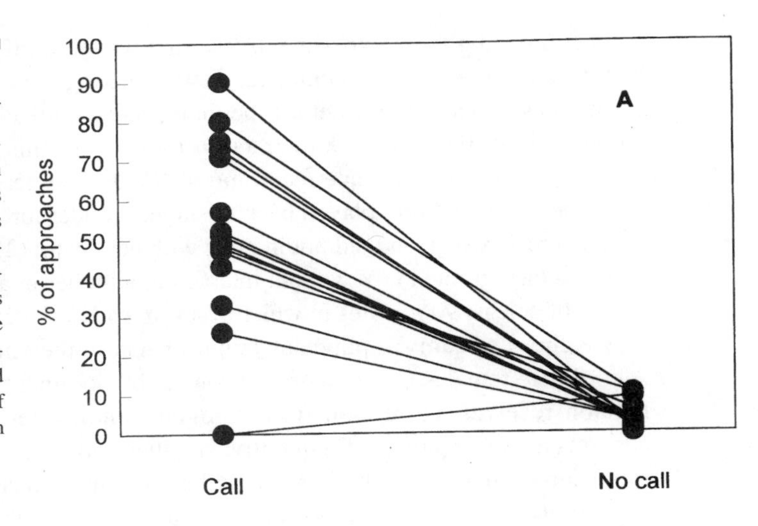

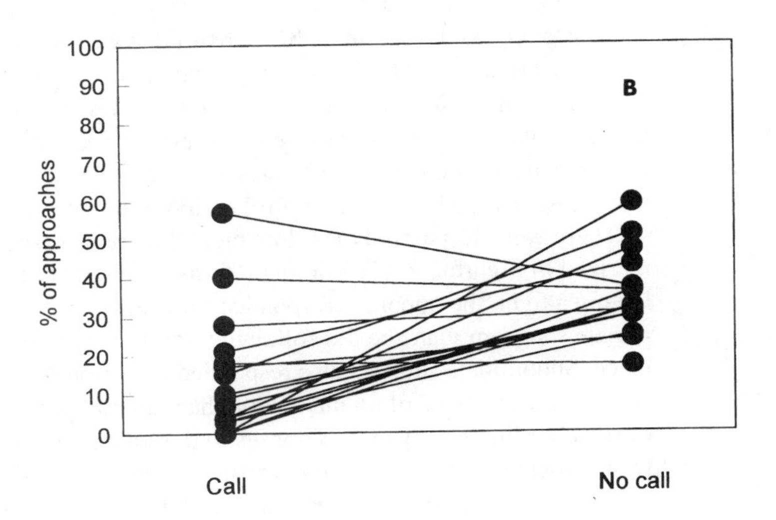

The effect of vocalizations on subsequent behavior by baboons. The mean proportion of 17 females' approaches toward subordinate partners that was followed either by (**A**) friendly behavior by the dominant or (**B**) supplanting of the subordinate. Approaches are divided according to whether the dominant female grunted as she approached or whether she remained silent. (Reprinted with permission of Academic Press Ltd, London, from Cheney et al., 1995, p. 251.)

Baboon females do sometimes grunt to one another after aggression. In an effort to examine the role of grunts in reconciling opponents, we carried out a number of systematic observations of aggressors and their victims. Whenever two females were involved in an aggressive interaction, we followed the aggressor for 10 minutes to determine whether she subsequently interacted with her victim in any way (Silk et al., in press). In 5% of 502 such samples, the aggressor subsequently interacted in a friendly manner with her opponent by touching her, grooming her, or interacting with her infant. Of these friendly interactions, 85% also included a grunt by the aggressor. In 9% of the situations, the aggressor grunted to her victim but did not interact in any other way.

These observations suggest that vocalizations alone, even in the absence of other affinitive interactions, may function to reconcile opponents. Nevertheless, the significance of the grunts themselves was difficult to assess simply from observations because grunts often occurred in conjunction with other friendly behavior, such as grooming or infant handling. To determine whether grunts might function to reconcile opponents even in the absence of other affinitive interactions, therefore, we designed a series of playback experiments.

In conducting these experiments, we first waited until a higher-ranking female, X, had threatened or chased an unrelated, lower-ranking female, Y. We then followed X for 10 minutes to determine whether she interacted affinitively with her opponent, and, if so, what form the affinitive interaction took. After this period, but within the next 30 minutes, we played a tape-recording of X's distress scream to Y and videotaped Y's response. Screams were played back to subjects under three conditions: (1) after X had been aggressive to Y and did not interact with her again; (2) after X had been aggressive to Y and then grunted to Y without interacting with her in any other way; and (3) after a period of at least 90 minutes in which X and Y had not interacted.

We chose screams as playback stimuli because they mimicked a context in which subordinate females are sometimes attacked by dominant individuals. When a female baboon receives aggression from a higher-ranking female or male, she typically screams at her opponent. Frequently, she then redirects aggression by threatening a more subordinate individual. We hypothesized that if a subordinate female heard the scream of an unrelated, higher-ranking individual, the subordinate would interpret the call as a potential threat to herself (Cheney et al., 1995). We predicted that Y would react strongly to the sound of X's scream if X had recently threatened Y but had not reconciled (evidenced by X's grunting) with her. Y's response in this context should be stronger than it was following a control period when the two females had not interacted. If, however, X had grunted to Y after threatening her, Y's anxiety should be diminished. We predicted that Y's response after vocal reconciliation should be similar to her response following the control period of no interaction.

There were 15 pairs (dyads) that met all three test conditions. The results are summarized in Figure 2.2. If a dominant female grunted to her subordinate opponent following a fight, the opponent responded for a significantly shorter period of time to that female's scream than she did following a fight when no further interaction had taken place. Subordinate subjects also responded less strongly to dominant females' screams after a control period of no interaction than after a fight with no reconciliation. In contrast, subordinate subjects' responses to dominant females' screams following a fight with a vocal reconciliation were statistically indistinguishable from the responses following a control period of no interaction. A theory of mind or not?

Observations and experiments, therefore, both suggest that vocalizations function in baboon society much like conversations function in human society: to reconcile, to appease, and to restore the relationships of rivals to their former baseline levels. But before we conclude that grunts are, in this functional sense, linguistic, we must look more closely at their underlying mechanisms.

There are at least two fundamentally different processes by which grunts might achieve their function of reconciliation and appeasement. On one hand, baboons may have a theory of mind, and when they vocalize, individuals may attribute mental states different from their own to their audience. Dominant females, if this were true, would grunt to alleviate their opponent's anxiety and to reassure the opponents that they are no longer angry. This would be a very "linguistic" human explanation.

On the other hand, baboons may *lack* a theory of mind, and, rather than attributing thoughts and beliefs to one another, callers and listeners may be responding to different behavioral contingencies (If she does such-and-such, I can relax; if she does so-and-so, I can't). Dominant females, in this case, would grunt to their victims because they are in a friendly mood and want to interact with their opponents' infants, and subordinate females would have learned, through experience, that grunts signal a low probability of attack. As a result, the subordinates would be less anxious when a dominant grunts to

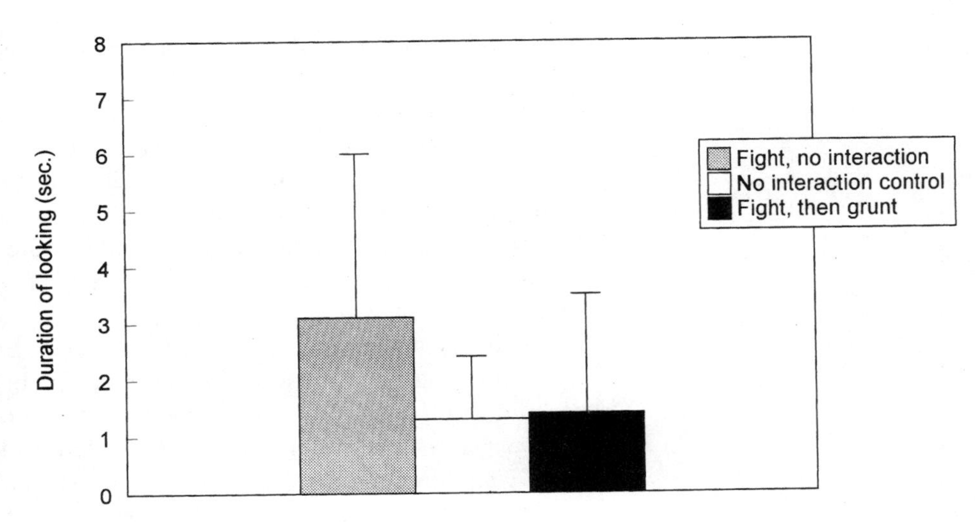

FIGURE 2.2

The strength of subjects' response to playback of a dominant animal's scream under three conditions. Vertical axis shows the duration of subjects' responses to the screams of dominant opponents after (1) the dominant animal threatened the subject and did not interact with her again; (2) the two females had not interacted for at least 90 minutes; and (3) the dominant animal threatened the subject and then reconciled by grunting to her. Histograms show means and standard deviations for 15 dyads in each of the three conditions. Subjects' responses were scored as looking in the direction of the speaker.

them, and grunts would function to reconcile and appease. In contrast with the linguistic interpretation, this perfectly reasonable explanation requires none of the sophisticated attribution of mental states that underlies human language.

Distinguishing between the two hypotheses is crucially important to any discussion concerned with the evolution of language. Each hypothesis is consistent with what we know about the function of baboon grunts — they appease, and they lead to reconciliation — but one hypothesis describes communication that is based on simple traditional principles of animal learning; the other describes communication that is infinitely more complex and genuinely similar to language.

Other Species

Can animals distinguish between their own beliefs and the knowledge and beliefs of others? Do individuals in any species, for example, ever take special steps to inform another who is ignorant or to correct another's false belief?

Hints that animals might be sensitive to the mental states of others come from recent work on the **audience effect** in various birds and mammals. When Peter Marler and his colleagues repeated Tinbergen's classic experiment by flying a model hawk over a group of chickens, they found that roosters adjusted their alarm-calling depending on who was nearby. For example, roosters gave alarm calls in the presence of a hen but remained silent when they were alone or with a female of another species (Gyger et al., 1986). In the wild, adult female ground squirrels are more likely to give alarm calls to predators if they have close kin present in the group than if they do not, but captive female vervet monkeys give more alarm calls to predators when they are paired with their own offspring than when they are paired with an unrelated animal of the same age and sex.

Although animals are clearly sensitive to the presence (or lack) of an audience, this fact does not prove that they are also sensitive to their audience's state of mind. In fact, there is very little evidence that monkeys or other animals ever take into account their audience's mental states when calling to one another. For example, alarm-calling by many birds and mammals is not "hard-wired" but depends on the presence of an appropriate audience. Individuals often fail to give alarm calls when there is no functional advantage to be gained by alerting others; for instance, when they are alone or in the presence of unrelated individuals (as observed for ground squirrels by Sherman, 1977; for downy woodpeckers by Sullivan, 1985; for vervet monkeys by Cheney and Seyfarth, 1985; and for roosters by Gyger et al., 1986). However, although the "audience effect" clearly requires that a signaler monitor the presence and behavior of group companions, it does not require the signaler to distinguish between ignorance and knowledge on the part of the audience. Indeed, in all species studied thus far, signalers call regardless of whether their audience is already aware of danger. Vervet monkeys, for example, continue to give alarm calls long after everyone in their group has seen the predator and retreated to safety.

Experiments with captive rhesus (*Macaca mulatta*) and Japanese macaques have found that mothers do not alter their alarm-calling behavior according to the mental states of their offspring. When given the opportunity to alert ignorant offspring of potential danger, mothers did not change their alarm-calling behavior (Cheney and Seyfarth, 1990a). Similarly, if vervet monkeys attributed mental states different from their own to others, they might be expected to correct or instruct their offspring in the appropriate use of alarm calls. This they never do. Infant vervets give eagle alarm calls to many bird species, such as pigeons, that pose no danger to them. Adult vervets, however, never correct their offspring when they make inappropriate alarm calls, nor do they selectively reinforce them when they give alarm calls to real predators such as martial eagles. Instead, infant vervets seem to learn appropriate usage simply by observing adults (Seyfarth and Cheney, 1986).

There is no doubt that the alarm calls given by monkeys function to inform nearby listeners of quite specific sorts of danger, but function is not always an accurate indicator of the presence of a theory of mind. Indeed, a closer examination of the mechanisms that underlie alarm calling suggests that when animals warn one another, they generally fail to take into account their audience's mental states.

Contact Barks of Baboons

When moving through wooded areas, female and juvenile baboons often give loud barks that can be heard up to 500 m away (Byrne, 1981). The contact barks can potentially function to maintain group cohesion because, on hearing one or more barks, an individual that has lost contact with others knows immediately where some group members are.

Because contact barks from different animals often occur at roughly the same time, it seems, to a human observer, as if the baboons are answering one another. But can we really be sure? After all, many calls could occur at roughly the same time because, when the group is widely spread and on the move, many animals are afraid of getting left behind and lost.

One hypothesis based on the attribution of mental states predicts that a baboon will answer the contact barks of others even when the baboon is in the center of the group progression and is at no risk of becoming separated from others. On the other hand, if a baboon is incapable of understanding that other individuals' mental states can be different from its own, the baboon should be unable to recognize when others have become separated from the group if the baboon is not in the same situation — just as the children younger than 4 years were unable to recognize that Maxi's knowledge of the candy's location might be different from their own knowledge. To test between the hypotheses, we gathered data on the social context of the contact barks given by 23 adult female baboons during a 3-month period.

Analysis of almost 2000 contact barks showed that the calls were, as expected, highly clumped in time (Cheney et al., in press). Were females really answering one another? If females had given answering calls *at random*, 96% (22/23) of each individual's calls should have followed a call by another female, and 4% (1/23) should have occurred after one of her own calls. In fact, 22 of 23 females gave fewer contact barks in the 5 minutes after a contact bark by another female than would have been expected by chance. Even close kin failed to answer each other's contact barks more often than expected by chance. In contrast, all 23 females answered themselves at least ten times more than expected by chance.

These data argue against the hypothesis that calls were clumped in time because females were answering one another. Instead, it seems that clumping of calls occurred primarily because each female herself, when she called, was likely to give a number of calls, one after the other.

As a further test of the hypothesis that females were not answering one another but instead were giving barks depending primarily on their own position, we carried out a series of 36 playback experiments in which we played to subjects the contact bark of a close female relative (mother, daughter, or sister). In 19% of trials, subjects did in fact answer their relative's contact bark by giving at least one bark themselves within the next 5 minutes. In no situation did other, unrelated females in the vicinity respond to the playbacks with a call (Cheney et al., in press).

At first inspection, the results might be taken as weak evidence for the selective exchanging of contact barks among close kin. After all, 19% of playbacks elicited an answer from close kin, but, under normal conditions, barks were answered by another female only 7% of the time. However, closer examination reveals that subjects answered playbacks of their relatives' barks primarily when they themselves were peripheral and at risk of becoming separated from the group. As shown in Figure 2.3, subjects that were in the last third of the group progression were significantly more likely to answer their relatives' contact barks than were subjects that were in the first two-thirds. Similarly, they were significantly more likely to give answering barks when there was no other female within 25 m than if there were at least one other female nearby (see Figure 2.3). Subjects were also more likely to call when the group was moving rather than feeding, though not significantly likely.

Observations and experiments, therefore, both argue against a theory of mind as the mechanism underlying baboon vocalizations. Apparently, baboons do not give contact barks with the intent of sharing information, even though the calls may ultimately function to allow widely separated individuals to maintain contact with one another. Like the progression, contact, and food calls given by other species of primates, contact barks of baboons appear to reflect the signaler's own state and position rather than the state and position of others.

NONHUMAN PRIMATE VOCALIZATIONS AND HUMAN LANGUAGE — SIMILARITIES AND DIFFERENCES

The vocalizations of nonhuman primates share a number of similarities with human speech. Many calls given by vervet monkeys, for example, are functionally semantic and denote objects or events in the external world. Like humans, vervets compare and

FIGURE 2.3

The effect of social context on individuals' responses to playback of a contact call given by one of their close relatives. Vertical axis shows the percentage of subjects who answered their relatives' contact barks in the 5 minutes following playback, under three different conditions. Data are based on 36 trials involving 18 subjects. (Reprinted with permission of Academic Press Ltd, London, from Cheney et al., 1995, p. 254.)

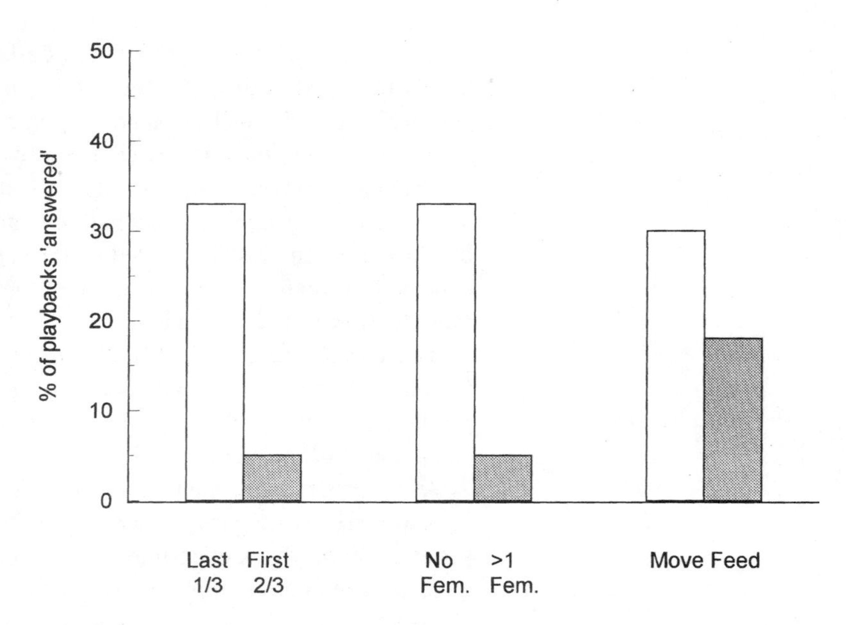

classify calls according to meaning, not just acoustic properties. Vervets judge some acoustically different calls to be the same if the calls refer to similar events. The monkeys behave, in other words, as if they recognize the referential relation between calls and the things for which they stand. Vocalizations are not just sounds, but sounds that represent something.

The calls given by monkeys during social interactions also appear to serve many of the same purposes that human speech has, in the sense that the calls seem to mediate social interactions, to appease, and to reconcile. Other calls function to inform individuals about the caller's location and to maintain group contact and cohesion.

Despite the functional similarities, however, the mental mechanisms underlying nonhuman primate vocalizations appear to be fundamentally different from the mechanisms underlying adult human speech. When calling to one another, monkeys seem to lack one of the essential requirements of language: the ability to take into account another's mental state.

Hypotheses based on the attribution of mental states — a theory of mind — make quite specific predictions about what we should see when we look closely at the vocal communication of nonhuman primates. In the context discussed here, an individual baboon that attributes mental states different from its own to others should adjust the alarm or intergroup calls according to her audience's knowledge, and the baboon should selectively inform ignorant individuals more than knowledgeable ones. The baboon should also correct offspring when they give alarm calls to inappropriate species. A dominant female baboon that attributes emotions to others should grunt to a subordinate victim to alleviate her victim's anxiety even though, being dominant, she feels no anxiety herself. Similarly, a baboon capable of attributing confusion or anxiety to others should answer another individual's contact barks regardless of its own position in the group progression.

Despite a variety of tests, however, there is no evidence that monkeys attribute mental states to one another. Monkeys appear not to call with the intent of providing information or influencing listeners' beliefs. Instead, listeners appear to respond to calls based on learned behavioral contingencies.

Although vervet monkeys, like many species of birds and other mammals, may vary their rates of alarm calling depending on the composition of their audience, they do not act deliberately to inform ignorant individuals more than knowledgeable ones (Cheney and Seyfarth, 1990a, 1990b). A vervet's alarm call alerts other animals to the presence of danger, and callers call regardless of whether their audience already knows what is happening. Similarly, dominant baboons grunt to subordinates after a fight, not because they recognize that the subordinates are anxious, but because they want to interact in a friendly way with the subordinates, usually because the subordinate animal involved has a young infant (Cheney et al., 1995; Silk et al., in press). Baboons give contact barks when they are at the group's periphery and at risk of becoming separated from others. The primary factor that determines whether an individual calls is that individual's own condition — not the perception of the condition of others. Through experience, listeners learn that they can maintain contact with at least a subset of the group simply by listening to other individuals' calls.

In each situation, listeners are able to extract crucial information from a vocalization based on their own experiences. But their responses — adaptive though they may be — need not take any account of the signaler's mental states. Indeed, in each case, the meaning and function of calls are to a large part determined by the listener rather than the signaler. On hearing a vervet's intergroup *wrr*, a listener deduces that another group is nearby, and this representation allows the vervet to ignore any subsequent intergroup vocalizations, even those with different acoustic properties. On hearing a dominant baboon's grunt, a subordinate baboon deduces there will be no attack. On hearing another baboon's contact bark, a listener deduces the group's location and direction of travel. In each case, the listener extracts rich, semantic information from a signaler who may not, in the human sense, have intended to provide such information.

From the listener's perspective, therefore, nonhuman primate vocalizations share many similarities with the words that make up human language. Not only do calls function to inform others of specific features of the environment and the signaler's emotions and intentions, but they also appear to be judged and classified according to the representations which they instantiate in the listener's mind.

From the signaler's perspective, however, there are striking discontinuities between nonhuman primate vocalizations and human language, at least as it manifests in adults. The discontinuities are based not so much on the formal properties of the calls themselves than on the mental mechanisms underlying call production. In marked contrast to adult human language, the calls of monkeys do not seem to take into account listeners' mental states. As a result, monkeys cannot communicate with the intent of appeasing others who are anxious or in informing others who are ignorant.

There is no doubt that the vocal communication of nonhuman primates mediates complex social relationships and results in the transfer of very specific sorts of information. Equally clearly, nonhuman primate vocalizations affect listeners' mental states, in the sense that they change what other individuals know about the world and affect what they are likely to do. Compared with human language, however, the vocalizations of monkeys achieve this end almost by accident, without individuals being aware of the features of the system in which they are participating. Monkeys, and perhaps also apes, are skilled at monitoring each other's behavior. There is little evidence, however, that they are equally adept at monitoring each other's states of mind. A challenge for the future is to identify the selective factors that may have favored the evolution of a theory of mind in the communication and behavior of our early ancestors.

REFERENCES

Aureli, F. 1992. Post-conflict behavior among wild long-tailed macaques (*Macaca fascicularis*). *Behav. Ecol. Sociobiol. 31*: 329–337.

Aureli, F., van Schaik, C., and van Hooff, J.A.R.A.M. 1989. Functional aspects of reconciliation among captive long-tailed macaques (*Macaca fascicularis*). *Am. J. Primatol. 19*: 39–51.

Bauers, K.A. 1993. A functional analysis of staccato grunt vocalizations in the stumptailed macaque (*Macaca arctoides*). *Ethology 94*:147–161.

Boinski, S. 1991. The coordination of spatial position: A field study of the vocal behavior of adult female squirrel monkeys. *Anim. Behav. 41*: 89–102.

Boinski, S. 1993. Vocal coordination of troop movement among white-faced capuchin monkeys, *Cebus capucinus*. *Am. J. Primatol. 30*: 85–100.

Byrne, R.W. 1981. Distance vocalizations of Guinea baboons (*Papio papio*) in Senegal: An analysis of function. *Behavior 78*: 283–313.

Chapman, C.A. and Levebre, L. 1990. Manipulating foraging group size: Spider monkey food calls at fruiting trees. *Anim. Behav. 39*: 891–896.

Cheney, D.L. 1981. Intergroup encounters among free-ranging vervet monkeys. *Folia Primatol. 35*: 124–146.

Cheney, D.L. and Seyfarth, R.M. 1985. Vervet monkey alarm calls: Manipulation through shared information? *Behavior 93*: 150–166.

Cheney, D.L. and Seyfarth, R.M. 1988. Assessment of meaning and the detection of unreliable signals in vervet monkeys. *Anim. Behav. 36*: 477–486.

Cheney, D.L. and Seyfarth, R.M. 1989. Redirected aggression and reconciliation among vervet monkeys, *Cercopithecus aethiops*. *Behavior 110*: 258–275.

Cheney, D.L. and Seyfarth, R.M. 1990a. Attending to behavior versus attending to knowledge: Examining monkeys' attribution of mental states. *Anim. Behav. 40*: 742–753.

Cheney, D.L. and Seyfarth, R.M. 1990b. *How Monkeys See the World: Inside the Mind of Another Species* (Chicago: Univ. Chicago Press), 377 pp.

Cheney, D.L., Seyfarth, R.M., and Palombit, R.A. In press. Function and mechanisms underlying baboon contact barks. *Anim. Behav.*

Cheney, D.L., Seyfarth, R.M., and Silk, J.B. 1995. The role of grunts in reconciling opponents and facilitating interactions among adult female baboons. *Anim. Behav. 50*: 249–257.

Clark, A.P. and Wrangham, R.W. 1994. Chimpanzee arrival pant-hoots: Do they signify food or status? *Int. J. Primatol. 15*: 185–205.

Cords, M. 1992. Post-conflict reunions and reconciliation in long-tailed macaques. *Anim. Behav. 44*: 57–61.

Cords, M. 1993. On operationally defining reconciliation. *Am. J. Primatol. 29*: 255–267.

de Waal, F.B.M. 1989. *Peacemaking Among Primates* (Cambridge: Harvard Univ. Press), 294 pp.

de Waal, F.B.M. and van Roosmalen, A. 1979. Reconciliation and consolation among chimpanzees. *Behav. Ecol. Sociobiol. 5*: 55–66.

de Waal, F.B.M. and Yoshihara, D. 1983. Reconciliation and redirected affection in rhesus monkeys. *Behavior 85*: 224–241.

Gyger, M., Karakashian, S.J., and Marler, P. 1986. Avian alarm-calling: Is there an audience effect? *Anim. Behav. 34*: 1570–1572.

Hauser, M.D. and Marler, P. 1993. Food-associated calls in rhesus macaques (*Macaca mulatta*). II. Costs and benefits of call production and suppression. *Behav. Ecol. 4*: 206–212.

Hausfater, G., Altmann, J., and Altmann, S. 1982. Long-term consistency of dominance relations in baboons. *Science 217*: 752–755.

Henkin, L.J. 1940. *Darwinism in the English Novel, 1860–1910* (New York: Corporate Press), 303 pp.

Hiraldo, F., Heredia, B., and Alonso, J.C. 1993. Communal roosting of wintering red kites, *Milvus milvus*: Social feeding strategies for the exploitation of food resources. *Ethology 93*: 117–124.

Hockett, C.F. 1960. Logical considerations in the study of animal communication. In: Lanyon, W.E. (Ed.), *Animal Sounds and Communication* (Washington, D.C.: Amer. Inst. Biol. Sci.), pp. 292–340.

Jackendoff, R. 1994. *Patterns in the Mind* (New York: Basic Books), 246 pp.

Janson, H.W. 1952. *Apes and Ape Lore in the Middle Ages and the Renaissance* (London: Warburg Inst.), 384 pp.

Judge, P.D. 1991. Dyadic and triadic reconciliation in pigtail macaques (*Macaca nemestrina*). *Am. J. Primatol. 23*: 225–237.

Masataka, N. 1989. Motivational referents of contact calls in Japanese macaques. *Ethology 80*: 265–273.

Matsuzawa, T. 1985. Color naming and classification in a chimpanzee (*Pan troglodytes*). *J. Hum. Evol. 14*: 283–291.

Mitani, J. and Nishida, T. 1993. Contexts and social correlates of long-distance calling by male chimpanzees. *Anim. Behav. 45*: 735–746.

Morton, E.S. 1977. On the occurrence and significance of motivation-structural rules in some bird and animal sounds. *Am. Nat. 111*: 855–869.

Peacock, T.L. 1817. *Melincourt* (London: Hookman, Baldwin, Cradock and Joy), 486 pp.

Pinker, S. 1994. *The Language Instinct: How the Mind Creates Language* (New York: William Morrow), 494 pp.

Povinelli, D.J. 1993. Reconstructing the evolution of mind. *Am. Psychol. 44*: 493–509.

Premack, D. 1976. *Intelligence in Ape and Man* (Hillsdale, NJ: Lawrence Erlbaum), 370 pp.

Premack, D. and Woodruff, G. 1978. Does the chimpanzee have a theory of mind? *Behav. Brain Sci. 1*: 515–526.

Prior, K.A. and Weatherhead, P.J. 1991. Turkey vultures foraging at experimental food patches: A test of information transfer at communal roosts. *Behav. Ecol. Sociobiol. 28*: 385–390.

Richner, H. and Marclay, C. 1991. Evolution of avian roosting behavior: A test of the information centre hypothesis and of a critical assumption. *Anim. Behav. 41*: 433–438.

Robinson, J.G. 1984. Syntactic structures in the vocalizations of wedge-capped capuchin monkeys, *Cebus nigrivittatus*. *Behavior 90*: 46–79.

Sakuro, O. 1989. Variability in contact calls between troops of Japanese macaques: A possible case of neutral evolution of animal culture. *Anim. Behav. 38*: 900–902.

Savage-Rumbaugh, E.S. 1986. *Ape Language: From Conditioned Response to Symbol* (New York: Columbia Univ. Press), 433 pp.

Seyfarth, R.M. 1976. Social relationships among adult female baboons. *Anim. Behav. 24*: 917–938.

Seyfarth, R.M. and Cheney, D.L. 1986. Vocal development in vervet monkeys. *Anim. Behav. 34*: 1640–1658.

Seyfarth, R.M. and Cheney, D.L. 1992. Meaning and mind in monkeys. *Sci. Am. 267*: 122–129.

Seyfarth, R.M., Cheney, D.L., and Marler, P. 1980a. Monkey responses to three different alarm calls: Evidence for predator classification and semantic communication. *Science 210*: 801–803.

Seyfarth, R.M., Cheney, D.L., and Marler, P. 1980b. Vervet monkey alarm calls: Semantic communication in a free-ranging primate. *Anim. Behav. 28*: 1070–1094.

Sherman, P.W. 1977. Nepotism and the evolution of alarm calls. *Science 197*: 1246–1253.

Smuts, B. and Nicolson, N. 1989. Reproduction in wild female baboons. *Am. J. Primatol. 19*: 229–246.

Sorabji, R. 1993. *Animal Minds and Human Morals: The Origins of the Western Debate* (New York: Cornell Univ. Press), 267 pp.

Struhsaker, T.T. 1967. Auditory communication among vervet monkeys (*Cercopithecus aethiops*). In: Altmann, S.A. (Ed.), *Social Communication Among Primates* (Chicago: Univ. Chicago Press), pp. 281–324.

Sullivan, K. 1985. Selective alarm-calling by downy woodpeckers in mixed-species flocks. *Auk 102*: 184–187.

van Hooff, J.A.R.A.M. 1994. Understanding chimpanzee understanding. In: Wrangham, R.W., McGrew, W.C., de Waal, F.B.M., and Heltne, P.G. (Eds.), *Chimpanzee Cultures* (Cambridge: Harvard Univ. Press), pp. 267–284.

Wilkinson, G.S. 1992. Information transfer at evening bat colonies. *Anim. Behav. 44*: 501–518.

Wimmer, H. and Perner, J. 1983. Beliefs about beliefs: Representation and constraining function of wrong beliefs in young children's understanding of deception. *Cognition 13*: 103–128.

Wrangham, R.W. 1977. Feeding behavior of chimpanzees in Gombe National Park, Tanzania. In: Clutton-Brock, T.H. (Ed.), *Primate Ecology* (New York: Academic Press), pp. 168–212.

York, A.D. and Rowell, T.E. 1988. Reconciliation following aggression in patas monkeys, *Erythrocebus patas*. *Anim. Behav. 36*: 502–509.

CHAPTER 3

WHY ARE WE AFRAID OF APES WITH LANGUAGE?

Sue Savage-Rumbaugh*

ANIMAL MINDS VERSUS HUMAN MINDS

Since the time of Aristotle (and probably before), scholars have debated the mental competence of animals. Views on this topic have ranged from those of **animists**, who believed in the immortality and transmigration of souls from animals to human beings, to **mechanists**, who claimed that both animals and humans were nothing but biological machines acting as circumstances programmed them, to **vitalists**, who held that a continuity of some fashion existed between animals and humans. Vitalism was given added legitimacy by the concept of the **Great Chain of Being**, according to which the Designer of the world created all forms of life in a finely graded scale from barely alive, to sentient, to intelligent, to wholly spiritual (Figure 3.1). Humans alone occupied the final rung of the ladder, and a human's ascent to spirituality was possible (but not ensured), a feat denied to other living creatures (Wise, 1995).

The basic sentiments of these positions are still echoed today. The sentiments have been "enlightened" by Descartes, who drew a distinction between the mind and the body of a human — the body behaves in a mechanistic manner akin to the behavior of animals, but the mind operates according to the principle of rational thought, a capacity denied to animals (Figure 3.2).

The idea of the Great Chain of Being had to be seriously revised following Darwin's insights (Figure 3.3). Evolution, it seemed, did not proceed in neat steps according to a grand design, nor were forms immutable and fixed for all time. Regardless of whether souls were transmutable, bodily forms clearly were, and they had evolved and changed over time. Since many other creatures had previously existed, after Darwin, a Designer

*Departments of Biology and Psychology, Georgia State University, Atlanta, GA 30303 USA

FIGURE 3.1

Prior to Darwin, many European intellectuals espoused the idea of a "Great Chain of Being." According to this doctrine, widely supported by the church, all living organisms were designed according to a graded plan by a devine Creator. This plan arranged living creatures along a *scala natura* from the "lowliest" to humans, the "highest." In addition, humans alone were said to possess the crucial property called "soul." (Painting done by Edward Hicks, 1833.)

had to be viewed as either having made mistakes or having been willing to leave the emergence of life forms to the whims of natural selection (Wise, 1996).

An additional critical insight shaping modern discussions of these issues is the wide recognition of the relevance of apes and the emerging awareness of the close similarity of many of their behaviors to behaviors of humans (Figure 3.4). In terms of the basic building blocks of the DNA code of life, researchers now know that apes are more closely related to humans than to monkeys. Thus, the dichotomy framed by the words *animal* and *human* is no longer valid.

These "new facts" have forced revision of Descartes' distinction between mind and matter (Sibley and Ahlquist, 1987). Mind is now acknowledged to be a creation of matter, and understanding how the mind–brain works is viewed as equivalent to understanding the neurobiology of brain function. However, unlike Descartes, modern neurobiologists do not equate understanding the mechanism with a mechanistic view of humans. Indeed, the new neurobiology avidly seeks the physiological basis of consciousness (Churchland, 1995). Several theories exist to explain the emergence of consciousness, but all neurobiologists see consciousness as an emergent property of a complex nervous system (Churchland, 1986).

The fact that science can now conceptualize a nonmechanistic phenomenon such as consciousness as arising from and being governed by the basic properties of neurologi-

FIGURE 3.2

Descartes contended that the body and mind of humans were functionally separate entities. The body followed the "mechanistic tendencies" thought to be reponsible for all animal activities. The mind, however, was thought to operate according to completely different principles, irrevocably separating the "mental world" of humans from that of animals. (Pen-and-ink drawings done by Shane Savage-Rumbaugh, 1996. Used with permission.)

FIGURE 3.3

With the emergence of Darwinian theory came the first insights that bodily forms changed through time and that humans and animals were made of similar materials. The "survival of the fittest" concept also challenged the view that life inevitably resulted in a progression toward "higher" forms. The barrier between the bodily form of humans and those of animals became understood as the result of a succession of chance events. (Pen-and-ink drawings done by Shane Savage-Rumbaugh, 1996. Used with permission.)

cal matter still leaves open the question of the human–animal relationship. There are differences, of course, between the brains of various species, but the more closely related the species, the less the difference. Between humans and apes, it is hard to find any difference, other than overall size, that would seem to account for what we perceive to be extremely large differences between the way human minds and the minds of animals analyze and process events. But entrenched views die slowly. Even though scientists are willing to assume that properties such as consciousness can emerge from complex neurological connections, many are not yet willing to accept a continuity between the minds of humans and the minds of apes (Pinker, 1994).

Thus, although there is no obvious neurological difference between humans and apes, a difference, nevertheless, must exist. Because the difference is not obvious, it must be small, but it is a difference that has a large effect. This difference, not yet clearly identified in terms of neural structure, has been labeled a **grammar module**. According to this view, all things that seem to set human beings apart from animals are assumed to be reducible to a single, unique capacity that finds the essence of its expression in what is called language. The capacity for language is presumed to undergird and make possible rational thought, and it is manifest in its clearest form in a human's grammar and in the concept of numbers. In essence, it is an ability to relate the structural components of abstract symbol systems to one another and to manipulate them according to rules.

FIGURE 3.4

The modern era has witnessed the "discovery" of apes by science and recognition that their behaviors are like human behaviors in ways that previous generations of scholars never imagined. Much criticism has befallen studies in which apes are reared as though they were human children, yet these studies have revealed the extent to which humans and apes share the capacity for learning and cultural transmission as a basis of behavior. Because much of social behavior is learned, that which seems "natural" tends to be that which is learned at an early age, often even before humans are aware that they are learning.

The "Holy Grail" of Grammar

Thus, the new view of human–animal relationships, enlightened by science and freed from religious dogma, is that humans are born with a sort of mathematical device inside the head (Bickerton, 1990). The device permits the person to relate rules for arranging and organizing symbols to real-world events in specific ways. So efficient is the encoding that the rules do not take up much space, nor do they require any dramatic restructuring of the brain. A human child need only be brought up with exposure to the activities of other humans to enable the human endowment to flower. Animals, being devoid of language, are said to lack the module of innate intelligence. Because they lack this analyzing capacity, animals are viewed as confined to learning complex chains of actions. Though the chains can become extremely elaborate, they remain, nonetheless, dependent on stimuli from the environment. According to this view, animals can never free themselves from their surroundings and contemplate the existence of self and nature as can human beings (Chomsky, 1988).

Even though modern neuroscience has discarded the difference between mind and matter, it has nonetheless retained the view that there exists a fundamental, functional difference between the minds of humans and the minds of animals. By positing the existence of a bit of critical (though unidentifiable) neural tissue that is said to contain something akin to the alphabet of thought, neuroscience has managed to incorporate Descartes, Darwin, DNA, and Goodall in its cosmology, without altering the central basic tenet that humans alone are capable of conscious, rational thought.

Nonetheless, this mainstream view is not a completely comfortable one, as can be seen from the recent rise of issues about animal rights. The problem, it seems, is that humans can clearly see that many animals, particularly mammals, are like themselves in many ways. Other mammals appear to feel as humans do when they are sad, hungry, tired, lonely, frightened, or angry. Other mammals take steps recognizable by humans as steps that humans might take to deal with such circumstances. Many animals appear to be as emotionally attached to their young, their mates, and their homes as humans are. Many can do things that humans cannot do because of the limitations of human senses. Many appear to share complex communication systems that humans cannot decipher in any but the crudest fashion (McFarland, 1984), and many build complex shelters and modify objects to use as tools (Beck, 1980). Yet, in spite of recognizing all the similarities between animals and humans, humans wish to maintain that they are somehow different and superior. Why? Perhaps, at least in part, the answer is that animals are very useful to humans. They are important sources of food, clothing, and labor for our species. Humans are economically dependent on animals for well-being, and it is thus reassuring for humans to regard animals as different.

The fact that no animals speak is convenient. It allows claims that animals lack reason and souls to be accepted without challenge. Deprived of the gift of language, animals are the ideal disenfranchised group.

WASHOE — THE FIRST APE TO "SPEAK"

Having learned to hunt, eat, and domesticate many species of animals, a human's conceptual and legal separation of self from all "lesser" creatures would probably have remained more or less intact were it not for the fact that a chimpanzee named **Washoe** was reported, in 1971 by Allen and Beatrice Gardner, to be able to talk (Gardner and Gardner, 1971). Washoe could not actually speak but was nonetheless capable of using a human language. The language was American Sign Language (ASL) for the deaf, a language used by millions of hearing-impaired people. Washoe was never suggested to be fluent in the language, only to be able to express simple desires and needs. Nevertheless, this one report, at odds with assumptions regarding animals accepted by people since before the time of the Greeks, created a maelstrom of controversy (Wallman, 1992).

Because of the depth, the history, and the economic forces at play underlying a human's concept of self, of rationality, and of dominion over animals, it should come as no surprise that initial claims that apes could learn human language were met with skepticism and cries of fraud (Sebeok and Rosenthal, 1981). The first debates centered on attempts to determine whether what Washoe was doing was really using language. The debates clarified that although humans had been using language all the time, they had no clear understanding of what language was and no means to determine precisely whether Washoe's language was the same as human language.

Students of language had never before had to develop a consensus as to what was meant by *language*, because they had never before been challenged by the idea of an animal that possessed this ability. Of course, if Washoe had simply begun speaking in scholarly terms on the issue, no one would have contested her ability. But that did not happen. Washoe did sign, but her signs were often repetitive. They seemed to lack internal structure, and it was not always easy to determine what she was saying. This was the first time anyone had succeeded in teaching a human language to another species. Moreover, the articulations of ASL were not designed for a chimpanzee's hands, nor was Washoe's vocal tract fashioned for the articulation of human speech. But all new scientific endeavors are clumsy at first, and thus, there was no reason to conclude that Washoe's performance reflected *her* limitations rather than those of humans' ability to understand.

There was no doubt that Washoe could produce different signs when shown a wide variety of objects. She also generalized the signs to new objects of similar form or conceptual class. For example, she applied the sign for meat to hamburger, steak, chicken, and so forth (Gardner and Gardner, 1971). It soon became apparent that other chimpanzees could also designate objects and could do so for plastic tokens or geometric symbols. Was this *naming*? Some argued that this behavior was merely **paired-associate learning** (Terrace et al., 1979) and asserted that Washoe only knew which gesture to make when presented with a certain item, a behavior not the same as knowing that the sign was a name for an object.

It was not possible, of course, to ask Washoe whether she had learned paired associates or whether she really knew that her signs were symbols for objects. Evidently, the only way to determine how Washoe herself viewed her signs was to study how she used them in combination with one another. If her combinations exhibited a **syntax**, it would have to be assumed that Washoe had come to possess a grammar module similar to that of humans. But Washoe's signs were not made when she was alone. Rather, they were made only when humans talked to her, and her combinations often contained components of the combinations that her caretakers had employed. How much of what Washoe said was imitation of others (Seidenberg and Pettito, 1979), and how much reflected the mind of Washoe? Humans speak to be understood. If Washoe wanted a banana and she spoke to convey this, she also wanted to be understood. Were we not employing a double standard (Taylor, 1984) — allowing things to be "on" human minds, but not on Washoe's mind?

Language Is a Two-Way Street

An important difference between Washoe's use of language and human use was ignored in the early rush to determine whether Washoe was "speaking her mind." The difference lay not in what Washoe said, what she wanted, or what she did. Rather, it was in what she understood. Language, as humans use it, is a two-way street. Not only do humans tell others what is on their minds, but humans also listen to what is on others' minds. Humans listen and then show their understanding by their actions. It is the understanding and the actions that signal such awareness and reward those who talk to us. For instance, if someone says he wants a banana (or a hug, love, understanding, sympathy, or that he wants us to "change our opinion"), humans typically engage in some sort of action in response to the statement. In the simplest case, if the person asks for a banana and we have one, we might offer it. If we are selling it, we might offer it along with a comment regarding its cost, but offer it, nonetheless.

Here Washoe and human beings parted ways. Washoe was quite adept at *asking* for a banana, but completely inept at *giving* one. Indeed, Washoe was inept at listening and giving anything! Her language was that of acquisition only, of a speaker in a world of her own. For Washoe, language had only one function — making others do what she wanted. Half of human language was missing. Washoe showed little if any understanding of what was said to her. She had learned to speak but had not learned to listen.

Listening, it was thought, was simply hearing what someone else said. What was there to measure? Studies of child language done before the work with Washoe had examined what children *said*, not what they understood. Indeed, it was thought neither possible nor relevant to study what they understood. Understanding, if even considered, was thought to follow along with speaking.

The problem of understanding and responding was not limited to Washoe. The same problem characterized two other chimpanzees, **Sara** and **Lana**. In his earliest report on Sara, Premack (1976) described comprehension as something that predates language but dismissed the need for further study of comprehension. However, listening is complex because it requires the assumption of the perspective of others. It also requires that the listener assume that others have something to say that the listener does not know. In brief, listening requires what is called a **Theory of Mind**.

Apes Learn to Listen

The first realization that language understanding provided greater insight into the true nature and processes of language came when attempts were made to teach apes to convey symbolic information to one another (Savage-Rumbaugh, 1986). The initial attempts were carried out with two male chimpanzees, **Sherman** and **Austin**, who had learned lexical symbols (Figure 3.5). The study revealed an unrecognized deficit in Washoe's use of language. Sherman and Austin, like Washoe, could readily speak their minds, in the sense of expressing desires for bananas, soft drinks, candy, and so forth. They expressed their desires just as effectively as Washoe did. As long as the recipients of their requests were humans, their language appeared to function in a quite ordinary (if rather food-oriented) manner, just as Washoe's language had. Humans listened and either gave them the foods they asked for or refused their requests. When refused, Sherman and Austin asked for something else.

Nothing was recognized as unusual regarding their use of language until the two chimpanzees were asked to talk only with one another. Suddenly it became clear that language entailed much more than simply speaking what was on one's mind. If the two apes spoke their minds only to one another, not much happened, other than a shouting match. Without one party serving as a listener and attempting to understand and respond to the expressions of the other, language did not get very far. The experiments with Sherman and Austin made clear that to understand language one had to go far beyond the idea that words simply function to express the mind. It became clear that language is much more involved, for it coordinates behaviors between individuals by a complex process of exchanging behaviors that are punctuated by speech. Language functions in a tightly woven mesh of expectancies and constraints set up by patterns of reciprocal actions that serve as variations on a theme. Words mean much more than just what is on someone's mind.

The most important benefit resulting from attempts to study language in creatures other than humans was that it allowed researchers to begin to peer through the veil that for centuries had shrouded human thought about language. Suddenly researchers realized for the first time that language was not about expressing oneself, but about how humans structure their interactions with one another. Humans could free action from talk; humans could discuss what they were going to do, wanted to do, when they would

FIGURE 3.5

Sherman and Austin were the first apes to communicate with one another using a human-based symbol system. Here Sherman is shown handing Austin a piece of "money," which Austin will use to "purchase" food from a vending device. This work led to the realization that "symbol comprehension," "turn-taking," and "sharing" are all critical aspects of language acquisition, even more critical than the ability to name things.

do it, how to do it, and what they thought about doing it — all without acting. Humans could talk about acting before they acted, talk about it after they acted, and talk about actions that might never happen. Humans could do all these things only as long as words were linked to action in some manner at some specific time, and the nature of the linkage lay not in expressing feelings, but in acting in response to the expressions of others. Understood in this way, human language did not seem so far removed from the communication systems of other animals — except that it was assumed that language alone was learned behavior.

Recognition that the key to language lay not in expression, but in comprehension, permitted **Animal Language Research** to move away from the teaching/training paradigm. By focusing on what an ape understood, rather than on what the ape could say, researchers learned that language — even spoken human language — was not something that one had to condition in a species, nor was it a feat that was achieved only to earn a reward. Although Washoe, Sara, and Lana had shown that animals were capable of learning various pieces of the language pie, anyone who interacted with them was left with the unsettled feeling that the overall effect of the apes' capacities, although clearly intelligent and language-like, fell short of what actually happens when humans talk with one another.

However, after a time, Sherman and Austin actually talked to each other, and their behaviors were intertwined with their talking, just as they are for humans (Figure 3.6).

FIGURE 3.6

Although Sherman and Austin were taught to use lexigram symbols to communicate, it was found that this training was not essential to their ability to learn symbols. With no training they learned the symbolic logos on bottles and cans of familiar foods. In this series of drawings (made from video tapes), Sherman and Austin demonstrate that they not only recognize logos as symbols but can use them to communicate symbolic information to one another (Savage-Rumbaugh, 1986). (**A**) Sherman observes as a caretaker takes food and hides a portion of it in each of two containers. (**B**) Sherman is shown brand-name food logos to see whether he wishes to use them to tell Austin what is hidden in a container. Because Sherman has learned previously to communicate similar information, using lexigrams, he knows that if Austin can tell another caretaker what is hidden in the container, then he and Austin will be able to share the contents.

(**C**) Immediately, without training or trial-and-error, Sherman selects the Coke logo and hands it to Austin. Using his lexigram keyboard, Austin announces to his caretaker "See the Coke logo," indicating that Coke is the hidden food. (**D**) Sherman then passes the food container to Austin.

Sherman listened when Austin asked for things or for assistance, and Austin listened when Sherman asked. Each monitored the effectiveness of communication with the other and helped the other as needed. Thus, if Austin did not understand which food or tool Sherman needed, and so indicated by a puzzled expression, Sherman helped by pointing to the object. In so doing, Sherman linked action with symbol to make clear his intent. Certainly, Sherman and Austin did not discuss the nature of the world; their exchanges focused typically on their immediate needs. Yet they clearly understood the basic power of symbols to communicate things from their individual experiences to the mind of each other (Figure 3.7). There seems no other way to explain Sherman's behavior when he rushed inside to announce to people and to Austin "Scare outdoors!" when he saw a partially anesthetized ape moaning as it was transported past the building.

Events such as this are dismissed as anecdotal. Persons making such denials claim that these kinds of events should be reliably repeated. Yet if such events are repeated, the need for communication vanishes. Why should Sherman run in to communicate "scare outdoors" if a moaning chimp is carried past every day? Another argument for dismissing such events is to say the apes had been conditioned in some nonobvious manner, for example, Sherman was simply generalizing to the event from something he had been reinforced to do in the past. This explanation is equally unsatisfying when one realizes that attempts to purposefully condition behavior in this manner fail.

Explanations that trivialize the behavior and strip its meaning cannot account for the kinds of language use demonstrated by Sherman and Austin. Only if instances such as "scare outdoors" occurred frequently in contexts that were inappropriate could one conclude that the chimpanzees did not understand or intend to communicate as they did. Unfortunately, much early work with Washoe and Lana reported only interesting incidents, rather than all language usage. Thus, it seems possible that only appropriate utterances were selected and inappropriate ones were unreported. However, this was not true for Sherman and Austin — utterances such as "scare outdoors" were reported within a large and contextually complete body of information (Savage-Rumbaugh, 1986).

LANGUAGE IS NOT IN THE MIND — IT IS IN INTERACTION

Other apes were to take the understanding of the ape mind much further, and they did this not because they were more intelligent apes, but because of what had been learned from Lana, Washoe, Sara, Sherman, and Austin. It is often thought that the goal of animal language research is merely to determine the capacity of another species for language and that this ability is something that is easy to measure, like bone length or skull circumference. According to this view, measurements suggest a certain value; thus, language either exists or it does not. The problem with this approach is that language ability is not something that lends itself to measurement in the same way as does a physical property such as skull circumference. Language exists in the nature of interactions among individuals. The nature of the interactions depends on assumptions about what each individual brings to the interaction and on the history of that individual's interactions with other individuals.

What is measured, then, cannot be merely a function of what an ape has, but rather a function of what human beings do through interactions with apes and what humans comprehend to be the result of the interactions. This is why ape skills seem to change;

FIGURE 3.7

Sherman and Austin use their lexigram keyboard to talk to each other.

they depend on the person making the measurements and the way the measurements are taken. The measured abilities of apes differ because of the activities. The skills of an ape do not develop in a manner independent of the observer and the interactions of the observer with that ape, which does not mean that the apes are cued or that only "human-reared" apes are intelligent or capable of language. Rather, it means that language, as a capacity, cannot (and should not) be separated from the way in which it is used. Language is the sum of the ways in which it is used to coordinate behavior (Savage-Rumbaugh et al., 1993).

Syntactical constructions do not coordinate behavior. Rather, syntactical constructions make it easy to decode certain concatenations of symbols, but the syntactical unpacking is itself trivial. It follows simple rules that can be learned, as Premack's work with Sara so clearly demonstrated. Even a syntactical structure as complex as an "if-then" relationship is trivial, as a structure iself, to teach to a chimpanzee. What is not trivial is the myriad of ways that the structure can be used in different situations — an understanding of the complete contextual situation is required to make sense of the if-then structure.

KANZI — LEARNING LANGUAGE WITHOUT BEING TAUGHT

As has been documented (Savage-Rumbaugh and Lewin, 1994), this emerging view of language is especially well exemplified by work with two other chimpanzees, **Kanzi** (Figure 3.8) and his sister **Panbanisha** (Figure 3.9). The apes received no instruction in language. They were reared in a language-rich environment in which they were spoken to in the context of daily life, a context that included trips in the forest to locate food

FIGURE 3.8

Kanzi, a young bonobo (pygmy chimpmanzee), has just asked me (rather than his mother, who is walking behind) to carry him. Kanzi wants to go to a particular location, but when he clings to his mother, his hands are not free to point out his desires. Unlike his mother, I walk bipedally so my hands are free to support his weight and Kanzi can easily signal where he wants to go. Kanzi's pointing gestures appeared spontaneously when he was 1 year old. He tried to utilize them occasionally while being carried by his natural mother, Matata; however, she ignored him. So, he would ask me to carry him and then showed me where he wanted to go.

resources and general daily activities such as feeding, playing, grooming, cleaning, and watching television. The language used was English. In addition, throughout each day, colorful symbols on a **keyboard** were pointed to and the names of the symbols were spoken by the apes' caretakers (Figure 3.10).

These conditions are all that is needed for apes to acquire an understanding of spoken language equivalent to that of a 3-year-old child (Figure 3.11). The understanding includes past and future verb tenses and syntactical constructions that require embedding and word order. This is also all that is needed for apes to learn to discriminate among written symbols, to pair the symbols with spoken words, and to use the symbols spontaneously to communicate information about their desires and their thoughts.

Language comprehension far outpaces production in these circumstances. Four- to eight-word sentences are easily processed, even on the first occasion they are heard, but

FIGURE 3.9

Panbanisha, Kanzi's younger sister, sits by her keyboard and studies pictures of things and locations that she likes. Shortly after she was 1 year old, she began using such photographs to tell me where she wanted to go. Like Kanzi, she pointed, but for places far away she preferred to use photographs. During such journeys, she often would carry a picture of the destination she desired until it became clear that she was nearing the place depicted in the photograph.

construction of sentences is generally limited to a single word plus a gesture and sometimes a vocal sound. Construction is limited, in part by the time required to locate each symbol on the keyboard, because the process of finding each symbol interferes with the capacity to use multiple symbols in rapid sequence (Figure 3.12).

Realizing That Others "Think"

Without any special training, apes reared with exposure to spoken language and printed symbols are able to follow and easily participate in three- and four-way conversations — conversations that deal with the intentions and actions of multiple parties and with the states of mind of the parties (Figure 3.13 on page 62). Panbanisha can, for example, answer questions about what it is that another chimpanzee thinks is hidden in a box, even though she knows that what she has seen placed in the box is not what the other party thinks is there. Not only can Panbanisha understand and answer such questions appropriately the first time they are posed, but she is also able to understand and comment on the fact that a deceitful trick has been played if the contents of the box are switched when the other chimpanzee was not looking (Figure 3.14 on pages 64 and 65).

To answer questions such as "What does Liz [a caretaker] think is in the box?" Panbanisha not only must understand the syntactical structure of the sentence — realize,

FIGURE 3.10

Panbanisha found the keyboard to be a great object for both play and communication. From the age of 6 weeks, the keyboard was always near her. By the time she was 1 year old she was able to recognize many lexigrams.

FIGURE 3.11

Panzee, a common chimpanzee, was reared with Panbanisha to determine whether bonobos (pygmy chimpanzees) were unique in their ability to understand human speech without training. Both Panbanisha and Panzee spontaneously began to understand human speech and to pair lexical symbols with sounds. Like Kanzi, they needed no training to acquire this human communication system.

for example, that the verb *think* refers to the agent, Liz. But, she also has to understand that the sentence is not dealing with what is actually in the box, but, rather, with what it is that Liz perceives to be in the box and that Liz's perception is based on an earlier experience, not on the current contents of the box.

The capacity to understand that the perceptions and states of knowledge of different individuals may not accord, and that the word think can refer to the nonobservable perceptions of others, requires a complex understanding of the nature of individuals and their past and present experiences, which far surpasses the complexity of the syntactical structure of the sentence itself. The structure of a question such as "What does Liz think is in the box?" is similar to that used by Premack, namely, "What Mary give Sara?" To solve such a sentence, Sara needed only to select one token if Mary gave her a banana and another token if Mary gave her an apple.

However, to answer a question about what Liz thinks is in a box, Panbanisha needs a concept of what thinking is and a realization that thinking for others is not the same as her own experience of thinking. She must also understand and recognize events that have recently happened to determine what it is that Liz thinks at a given moment, and she must be able to differentiate events that have happened to her from events that have happened to Liz.

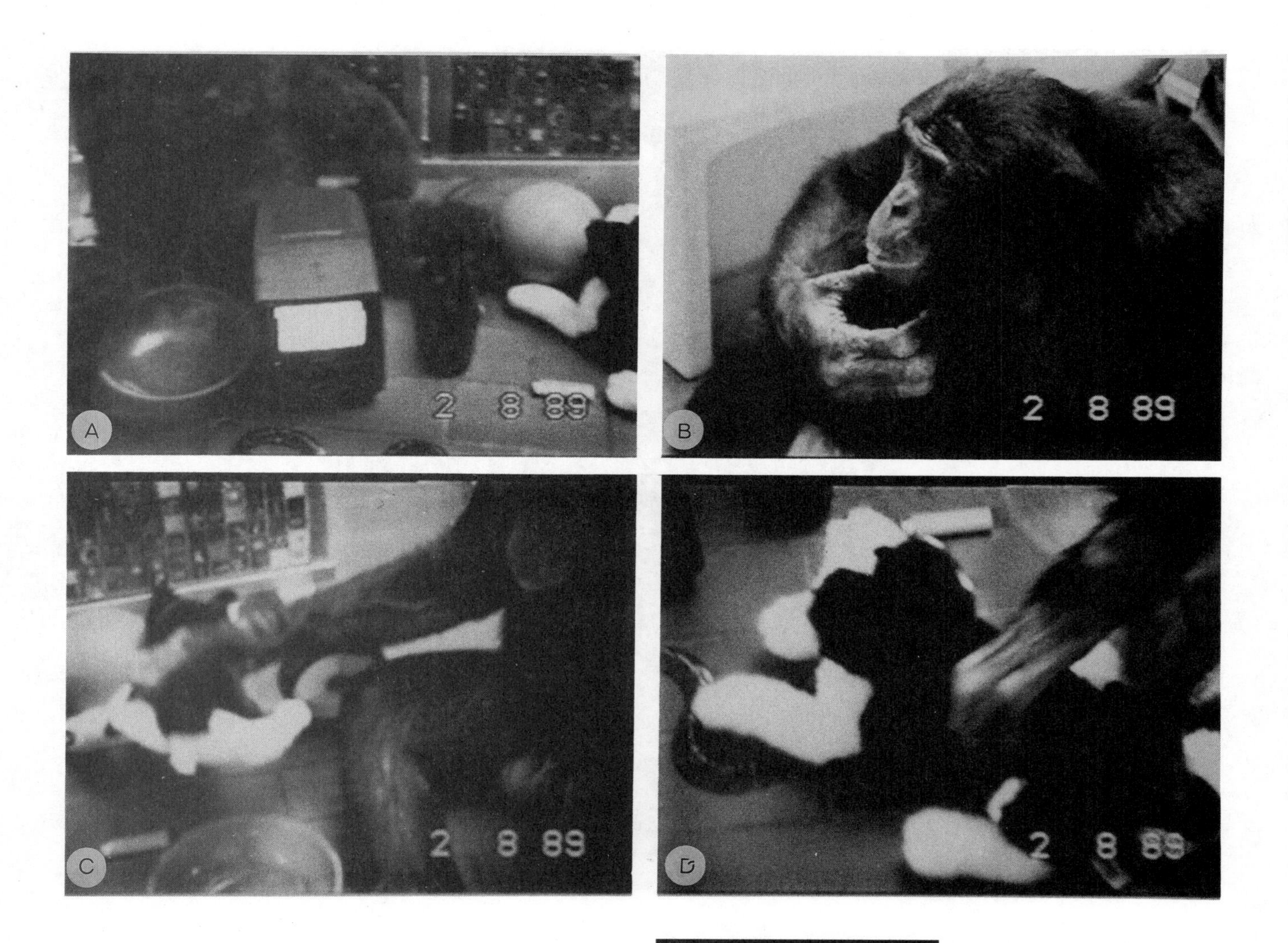

FIGURE 3.12

Kanzi not only understands individual words, but complex new sentences as well. To eliminate any possibility that his responses to sentences could have been learned, he was tested with more than 600 unusual examples. Such sentences included: "Can you feed your ball some tomato?" "Can you put the chicken in the potty?" "Can you make the snake scare Matata?" Kanzi's ability to understand these sentences compared favorably to those of a 2½-year-old child. This sequence of photos shows Kanzi's response to the question "Can you make the doggie bite the snake?" (**A**) The objects shown were in front of Kanzi at the time the question was presented. To keep the task from becoming trivial to Kanzi, sets of objects were changed after 2 to 5 sentences. (**B**) Kanzi listens to the question. It is important that Kanzi listen carefully because the experimenter speaking is behind a door, completely out of Kanzi's sight. (**C**) On hearing the sentence, Kanzi picks up the dog. (**D**) He moves the dog away from the other objects. (**E**) Kanzi selects the snake and looks at it. (**F–H**) Kanzi moves the snake toward the dog's mouth and pushes the mouth down on the snake with a biting action.

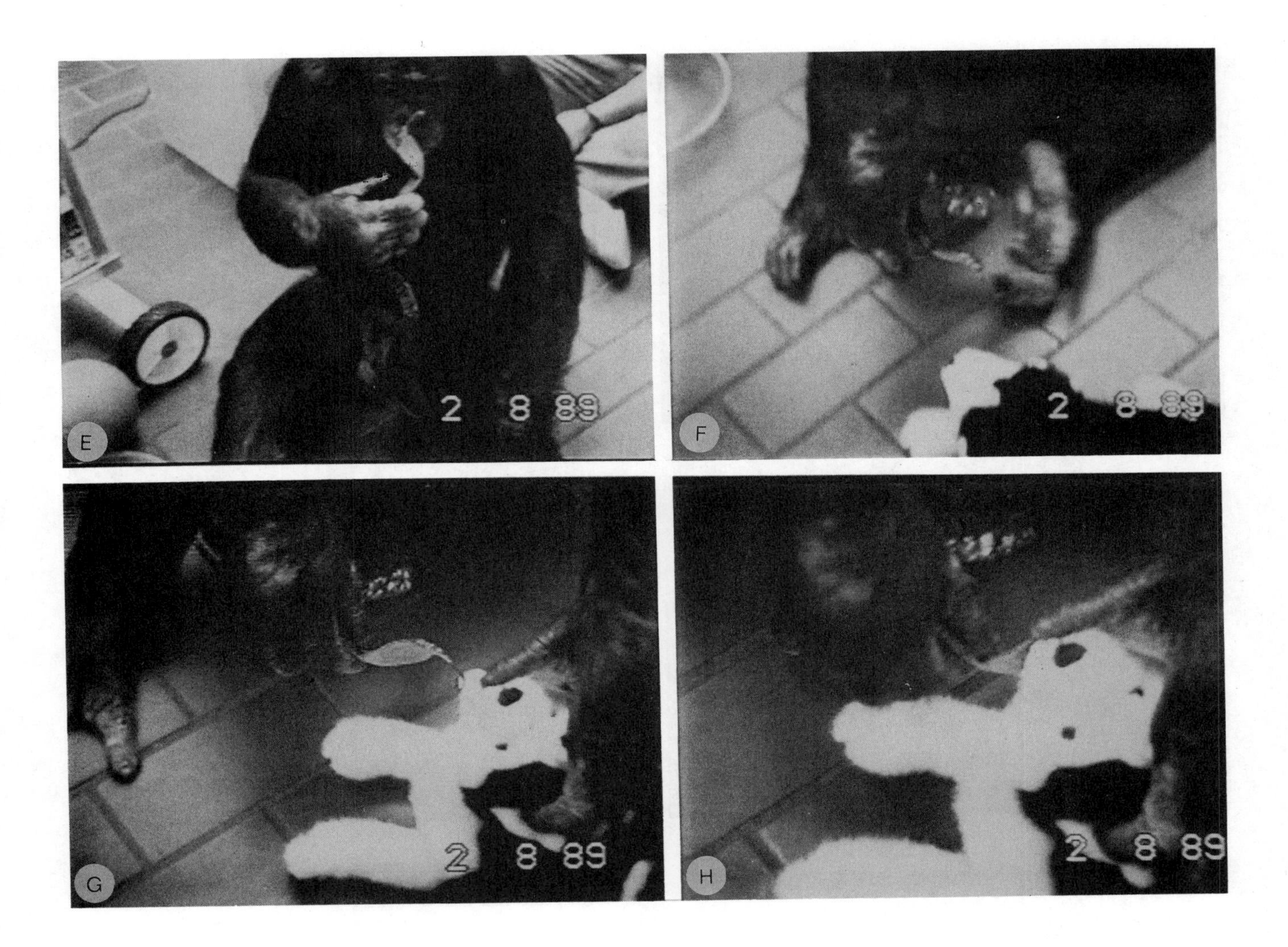

Panbanisha also needs to understand what has happened (specifically, that a substitution of an object has occurred and that a nonsubstitution expectancy has been generated for Liz) and that Liz herself is not aware of what has happened. Moreover, Panbanisha must realize that she needs to differentiate between Liz's knowledge and her own knowledge to answer the question posed. The ability to do this kind of computational analysis of a new situation and to reply symbolically and appropriately to such questions with new replies goes far beyond a syntactical understanding of which words in a sentence relate to one another as clauses, modifiers, and so on. No syntactical explanation of the question, the answer, or both could ever begin to characterize the true complexity of the language task in a situation such as this one.

Yet apes can do these things. They can understand all the components of such a situation, including the different participants in the situation, their different experiences, and their different perceptions. They also can understand linguistically posed questions about such a situation, even if it is new, and they can answer appropriately and even comment, giving their opinion about the propriety of the situation.

In the box example recounted above, Panbanisha told us that we were being "bad" to play such a trick on Liz. When the procedure was repeated with Liz's 4-year-old

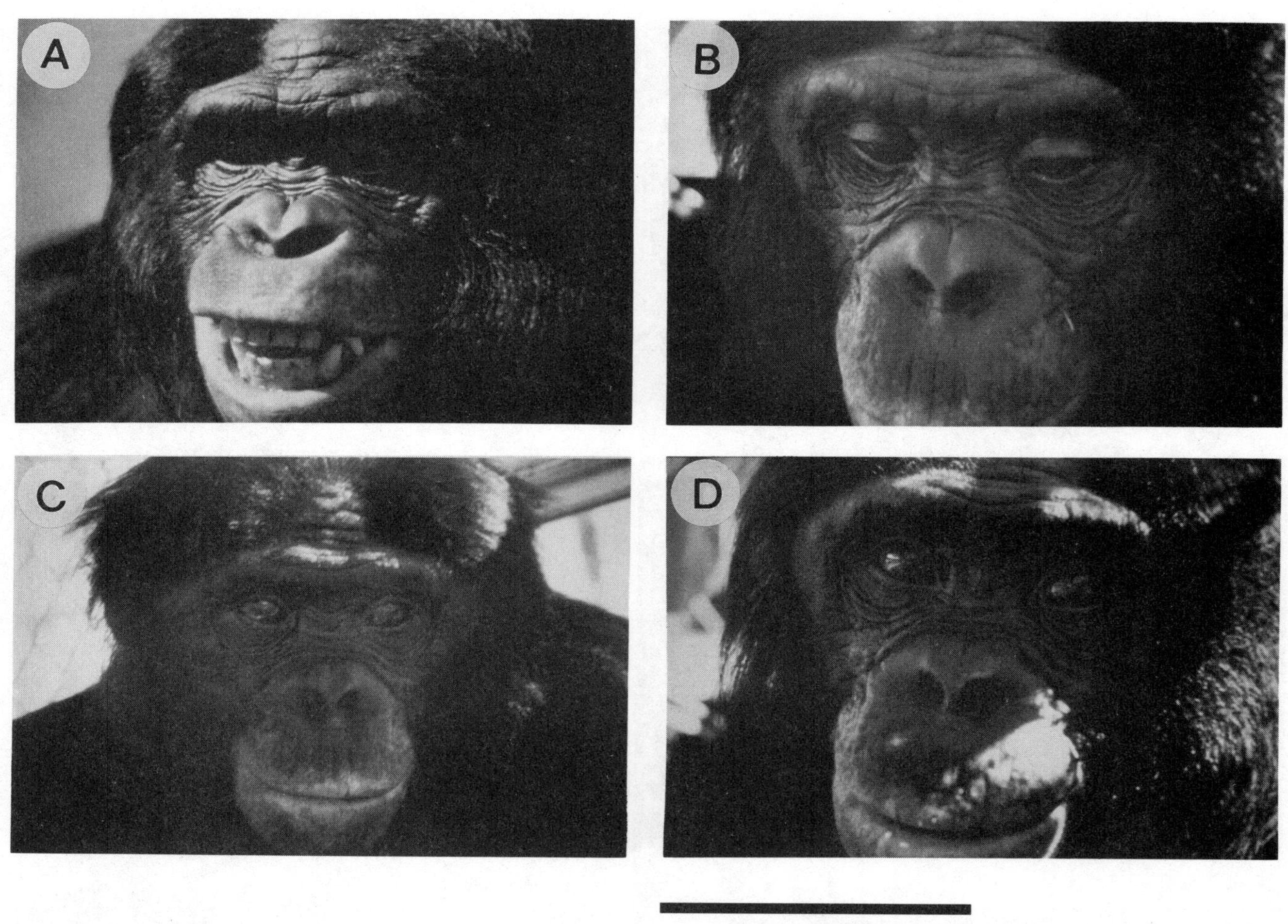

FIGURE 3.13

The faces of bonobos are very revealing of inner emotions, just like human faces. Here Kanzi is (**A**) happy, (**B**) reflective and thoughtful, (**C**) mildly uncertain, and (**D**) pensive.

daughter, Panbanisha's perceptions were similar; again she elected to tell us we were being "bad."

It is astonishing that apes that truly understand language can do all these things easily, even if their language production skills are typically limited to single symbols. Apes that do not understand language cannot do these things. We have no way of knowing, without language, why they fail. Do they not understand that different people can have different perceptions of events? Do they not know what is actually in the box? Do they not realize that Liz has thoughts about what is in the box? Do they know that things remain in boxes even when they are out of sight? We cannot even be certain the apes understand the task as something that has actually been "presented," that is, as a set of connected events in their environment that they are supposed to make sense of. We do not know whether the apes have viewed the set of activities (hiding something in a box, taking it out while Liz was not looking, replacing it with something else) as individual events, unconnected in any way, much less connected differently in Liz's mind than in their own minds.

Language leads humans to a consensual group interpretation that results in a psychological connectedness among such events. For humans, saying "Let's do this while Liz is not watching" sets up a connectedness between what Liz is now doing (that is, not watching) and what is now happening in the mind of an observer who understands the words. The connectedness makes possible inferences that might not otherwise occur. With apes that cannot understand sentences, researchers have no means of ascertaining what connectedness the apes assign to events, nor of querying them about their interpretation.

WINDOWS TO THE MINDS OF OTHERS

That one can query apes that understand language and that one does not need to train them to understand questions reveals that language functions to structure a joint perception of the world in ways fundamental to making possible joint interpretation and joint action. Such events are windows on the true power and value of language. Apes that have been exposed to language at an early age can peer through the windows in the same way humans do, which indicates that the mind of the ape cannot be very different from that of humans.

It no longer seems likely that communication among such creatures in nature can be accurately characterized as instinctive and devoid of meaning or intention. It also does not seem reasonable to think of such communication as nonlinguistic. It is probable that humans have focused so tightly on syntax and human speech that they have missed the true nature of language and, thus, failed to realize its existence in other species. Defining language in a peculiarly human fashion has glorified the form of language at the expense of understanding, even of denying, its function. The narrow view has led to the short-sighted conclusion that language is extant only in the human communication system and that communication systems of other creatures lack representation, structure, and meaning.

Animal language research has taken research a long way toward understanding who humans are and what language is. It has started to show that humans may have failed to recognize the intellect and power behind the communication that exists among members of other species.

TOWARD A SCIENCE OF ANIMAL CONSCIOUSNESS

Language is, indeed can be, nothing more and nothing less than the uses to which it is applied. People are particularly proud of the embedded and recursive structures that characterize the syntax of language, and many have assumed that such structures exist *only* in human language. Until one is willing to credit other creatures with the ability to learn to communicate complex messages, one can never discern similar syntactical complexities, should they exist. The fact that Kanzi and Panbanisha could so easily understand the use of syntactic devices reveals that these facets of our linguistic system are not beyond their reach and, thus, may exist in their natural communications (Savage-Rumbaugh et al., 1993).

FIGURE 3.14

Cartoons representing one version of a Theory of Mind test given to Panbanisha (discussion of this starts on page 57). In such tests, taken successfully by most 4-year-old children, the goal is to create a state of reality in the mind of a person other than the child being tested that differs from the reality the tested child knows to exist. Questioning is then used to determine whether the child comprehends that the knowledge of the other person differs from his or her own. Children who pass this test are said to exhibit a "theory of mind," that is, to understand that the minds of others are capable of registering information about their surroundings distinct from what they know to be correct. Children too young to pass this test are regarded as having an "egocentric" view of the world in which they assume that others know exactly what they know. (**A**) Two experimenters, Sue (*left*) and Liz (*right*), walk with Panbanisha (short figure, *center*). Liz asks Sue for candy, which Sue agrees to put in a box for Liz. (**B**) Liz watches as Sue starts to put candy into the box; then Liz departs, saying she will return soon. Panbanisha watches this event. (**C**) In Liz's absence, Panbanisha watches as Sue replaces the candy with a bug. (**D**) When Liz returns, Sue says "Here's your box" and gives the box to Liz. (**E**) Liz pretends to have difficulty opening the box. While she is trying to open it, Sue asks Panbanisha what Liz is looking for, and Panbanisha responds by pointing to the lexigram for candy on her keyboard.

Let's play a trick on Liz.
C
Here's your box.
D
What is Liz looking for?
It's hard to open
M&M's
E

Humans have been prohibited from seeing language skills in animals because they have assumed that any language must be like human language and must have encoded the kinds of things that humans have desired to encode. What humans need to realize is that animals may elect to classify the world differently than humans do. For example, humans may recognize a relationship between the movement of the Moon and the tide that animals may not, and humans may encode the relationship into language. Animals may sense a relationship between the movement of insects and the ripening of fruit in a remote location that humans may not, and animals may encode this relationship into their language. Because humans do not sense the relationship, the animals' encoding has no meaning — humans have tended to interpret that because animals do not point at and name objects as they do, animals have no language. What has been overlooked is that there are other sorts of things that are much more complicated and more important to convey linguistically, and that many of these other things are not "point-at-able." Given what we have now learned from apes, such as Kanzi and Panbanisha, humans can no longer afford to assume anthropocentrically that all other species lack language and conscious thought.

To achieve a science of animal consciousness, researchers must extend to animals the recognition that existence necessitates a living, thinking, experiencing progression of events. The nature of the experiencing can vary between species, between individuals, and between groups within a species. Understanding the nature and parameters of the variations is the essence of the study of animal consciousness. The goal should not be to understand *precisely* what it is like to be a bat, a dog, an ape, a mouse, a paramecium, or even another person. The goal must be to understand, partially, what it can be like to be any of these things and how the "beingness" of such entities changes with time, development, and circumstance.

The basic premise of the new paradigm must be that animal consciousness exists and that it is similar in some respects to the consciousness that humans experience, though it differs in degree and in its manner of manifestation in all living things. We must start with the premise that consciousness is an emergent characteristic of life, and we must move toward exploration and understanding of this characteristic in a myriad of life forms.

Once we acknowledge that consciousness is present in other animals besides ourselves, we can ask how and under what circumstances it manifests itself. What factors heighten or lessen the manifestation of consciousness? What is the extent of the boundary of consciousness under varying conditions? In attempting to answer such questions, we must first acknowledge that there are no final answers to the conditions of animal consciousness; there are only partial answers.

Persons who study animal consciousness can no longer assume the stance of a detached observational state. Instead, the scientist must aim to achieve some aspect of the participatory state, however inexact (Figure 3.15). Behavioral scientists must accept that they — like physicists — are a part of the frame of reference, and that it is inevitable that the consciousness of the observer affects the state of the system being observed. No two observers ever experience *precisely* the same perception of the consciousness of other beings.

Because all observers have different perspectives, the vast differentiation of life forms, both within and between species and within and through time, is possible. Science as a discipline must not resist what appears to be an imperfection or imprecision of method; scientists must devise new ways to study a more flexible, changing system whose parameters of existence reflect states of change, uniqueness, and perspective as joined aspects of observer and observed.

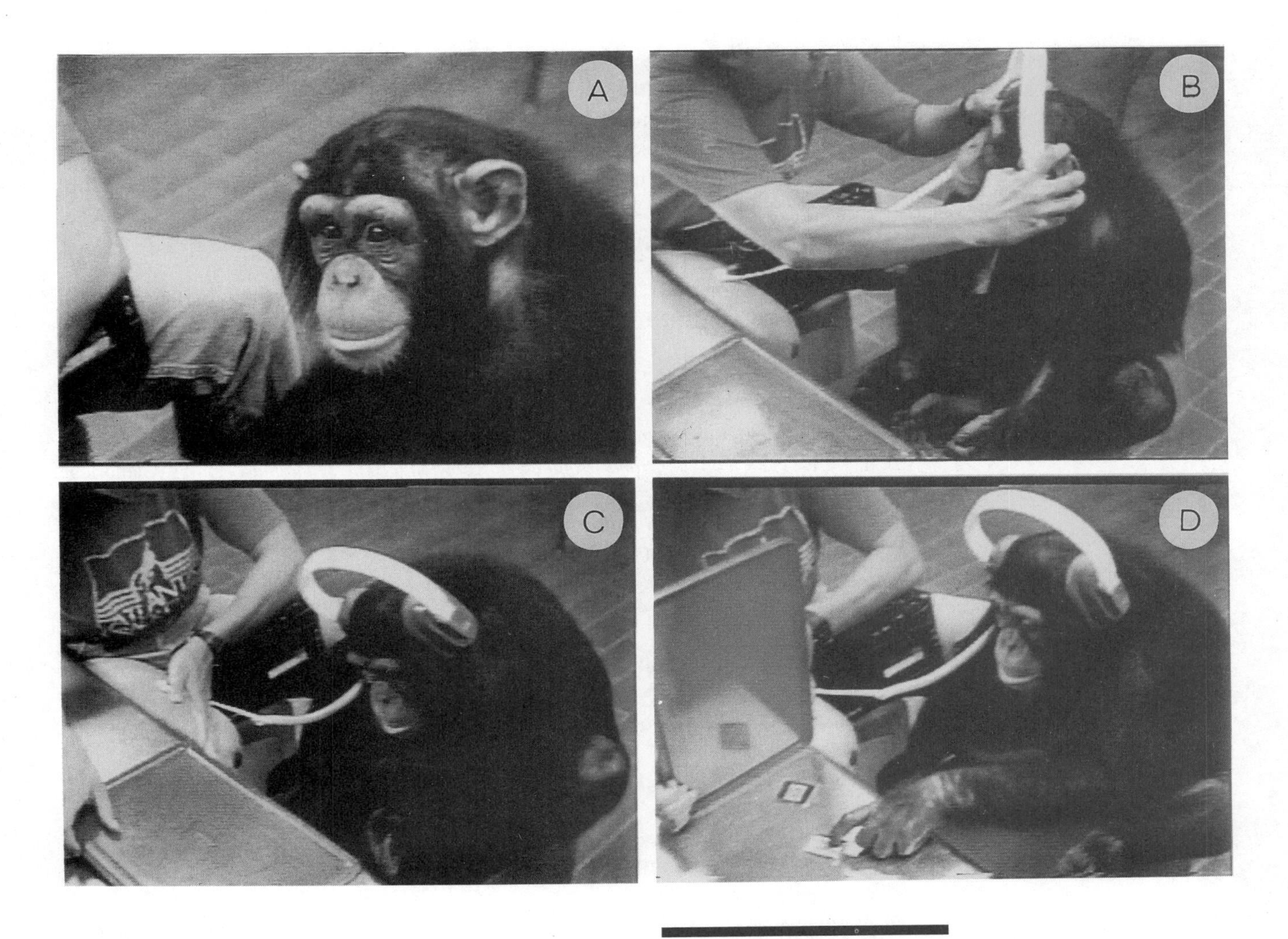

FIGURE 3.15

(**A**) Panzee participates in a test of English comprehension with a caretaker whom she knows well and trusts. (**B**) She permits headphones to be placed over her ears. (**C**) Panzee listens to the sound played from a tape recorder. (**D**) She points to one of three lexigrams to indicate the word she has heard. These types of tests permit a caretaker who is well known to an ape to ascertain the ape's capabilities and eliminate any chance that the caretaker's actions may inadvertently cue the ape to select the correct answer. Panzee needed no training and received no reward for participating in these tests; her relationship with the caretaker was the important factor motivating her to learn language and to participate in such basic tests of competency.

The methods of this new science remain to be developed, but the essence of what the methods can be lies in the situations that gave rise to the learning of a human language by the mind of a **bonobo** (pygmy chimpanzee). Language studies have not asked what is *in* a mind, but rather how experiences *form* minds. The experiences of Kanzi do not belong to him alone; they belong to all who have interacted with him. Their perceptions have shaped his character and state of knowledge, and he has shaped theirs. From the joint molding of mutual understanding have come many means of verifying that the understanding of one party is indeed valid for the other. The understanding and use of communication is central to achieving joint validation.

I believe it is time to change course. It is time to open our eyes, our ears, our minds, our hearts. It is time to look with a new and deeper vision, to listen with new and more sensitive ears. It is time to learn what animals are really saying to us and to each other.

REFERENCES

Beck, B.B. 1980. *Animal Tool Behavior: The Use and Manufacture of Tools by Animals* (New York: Garland STPM Press), 307 pp.

Bickerton, D. 1990. *Language and Species* (Chicago: Univ. Chicago Press), 297 pp.

Chomsky, N. 1988. *Language and Problems of Knowledge: The Managua Lectures* (Cambridge: MIT Press), 205 pp.

Churchland, P.S. 1986. *Neurophilosophy: Toward a Unified Science of the Mind/Brain* (Cambridge: MIT Press), 546 pp.

Churchland, P. 1995. *The Engine of Reason. The Seat of the Soul* (Cambridge: MIT Press), 329 pp.

Gardner, B.T. and Gardner, R.A. 1971. Two-way communication with an infant chimpanzee. In: Schrier, A.M. and Stollnitz, F. (Eds.), *Behavior of Non-Human Primates*, Vol. 4 (New York: Academic Press), pp. 117–184.

McFarland, D. 1984. *The Oxford Companion to Animal Behavior* (Oxford: Oxford Univ. Press), 657 pp.

Pinker, S. 1994. *The Language Instinct: How the Mind Creates Language* (New York: HarperCollins), 494 pp.

Premack, D. 1976. *Intelligence in Ape and Man* (Hillsdale, NJ: Erlbaum), 370 pp.

Savage-Rumbaugh, E.S. 1986. *Ape Language: From Conditioned Response to Symbol* (New York: Columbia Univ. Press), 433 pp.

Savage-Rumbaugh, E.S. and Lewin, R. 1994. *Kanzi: An Ape at the Brink of the Human Mind* (New York: Wiley), 299 pp.

Savage-Rumbaugh, E.S., Murphy, J., Sevcik, R., Brakke, K.E., Williams, S.L., and Rumbaugh, D.M. 1993. Language comprehension in ape and child. *Soc. Res. Child Devel. 58*(3-4): 1–256. Monograph.

Sebeok, T.A. and Rosenthal, R. 1981. The clever hands phenomenon: Communication with horses, whales, apes and people. *Ann. N.Y. Acad. Sci. 364*: 1–311.

Seidenberg, M.S. and Pettito, L.A. 1979. Signing behavior in apes: A critical review. *Cognition 7*: 177–215.

Sibley, C.G. and Ahlquist, J.E. 1987. DNA hybridization evidence of hominoid phylogeny: Results from an expanded data set. *J. Mol. Evol. 26*: 99–121.

Taylor, T. 1984. Linguistic origins: Brunner and Condillac on learning how to talk. *Lang. Commun. 4*(3): 209–224.

Terrace, H.S., Pettito, L.A., Sanders, R.J., and Bever, T.G. 1979. Can an ape create a sentence? *Science 206*: 891–900.

Wallman, J. 1992. *Aping Language* (New York: Cambridge Univ. Press), 191 pp.

Wise, S.M. 1995. How non-human animals were trapped in a nonexistent universe. *Anim. Law 15*: 1–54.

Wise, S.M. 1996. The legal thinghood of non-human animals. *Boston Coll. Environ. Affairs Law Rev. 23*: 471–546.

FURTHER READING

Corbey, R. and Theunissen, B. 1995. *Ape, Man, Apeman: Changing Views Since 1600* (The Netherlands: Prehistory Dept., Leiden Univ.), 408 pp.

Rogoff, B. 1990. *Apprenticeship in Thinking* (New York: Oxford Univ. Press), 242 pp.

Rumbaugh, D.M. 1977. *Language Learning by a Chimpanzee* (New York: Academic Press), 312 pp.

Savage-Rumbaugh, E.S. and Lewin, R. 1994. *Kanzi: An Ape at the Brink of Human Mind* (New York: Wiley), 299 pp.

Tuttle, R.H. 1986. *Apes of the World: Their Social Behavior, Communication, Mentality and Ecology* (Park Ridge, NJ: Noyes), 421 pp.

Wrangham, R.W., McGrew, W.C., De Waal, F.B.M., and Heltne, P.G. 1994. *Chimpanzee Cultures* (Cambridge: Harvard Univ. Press), 424 pp.

CHAPTER 4

THE MODULAR NATURE OF HUMAN INTELLIGENCE

Leda Cosmides* and John Tooby*

EVOLUTIONARY PSYCHOLOGY

The goal of research in evolutionary psychology is to discover and understand the design of the human mind. Evolutionary psychology is an approach to psychology, in which knowledge and principles from evolutionary biology and human evolutionary history are put to use in discovering the structure of the human mind. Evolutionary psychology is not a specific subfield of psychology, such as the study of vision, reasoning, or social behavior; it is a way of thinking about psychology, which can be applied to any area of human behavior or competence.

In this view, the mind is a set of information-processing procedures (cognitive programs) that are embodied in the neural circuitry of the brain. Realizing that the function of the brain is information-processing has allowed cognitive scientists to resolve at least one version of the mind–body problem. For cognitive scientists, *brain* and *mind* are terms that refer to the same system, which can be described in two complementary ways — in terms of its physical properties (the brain) or in terms of its information-processing operation (the mind). Described in computational terms, the mind is what the brain does. The physical organization of the brain evolved because physical organization brought about certain adaptive information-processing relationships. These organic computer programs were designed by natural selection to solve adaptive problems faced by our hunter-gatherer ancestors and to regulate behavior so that adaptive problems were successfully addressed. This way of thinking about the brain, mind, and behavior is changing how scientists approach old topics and is also opening up new ones. This chapter is an introduction to the central ideas and research strategies that

*Center for Evolutionary Psychology, University of California, Santa Barbara, CA 93106 USA

have animated evolutionary psychologists and to some of the common misconceptions that people often have about the field.

DEBAUCHING THE MIND — EVOLUTIONARY PSYCHOLOGY'S PAST AND PRESENT

In the final pages of *On the Origin of Species*, after Darwin had presented the theory of evolution by natural selection, he made a bold prediction: "In the distant future I see open fields for far more important researches. Psychology will be based on a new foundation . . ." Thirty years later, William James (1890) began to outline what evolution meant for the study of psychology in his seminal book, *Principles of Psychology*, one of the founding works in experimental psychology. James talked a great deal about *instincts*, a term used to refer, roughly, to specialized neural circuits that (1) are common to every normal member of a species; (2) are the product of the species' evolutionary history; and (3) are acquired by a species' particular program structure because a particular set of rules solved an adaptive problem for the organism. (For example, if one is without defenses and is chased by a predator, one runs away.) These instincts have been referred to by various terms — modules, cognitive programs, adaptive specializations, evolved circuits, innate procedures, mental adaptations, evolved mechanisms, natural competences, and so on. The thousands of evolved circuits in our own species constitute a scientific definition of *human nature* — the uniform architecture of the human mind and brain that reliably develops in every normal human just as do eyes, fingers, arms, a heart, and so on.

It was common during James' time, as it is today, to think that other animals are ruled by instinct but that humans have lost all or almost all their instincts and have come to be governed instead by reason and learning. According to this view, the replacement of instinct with reason is why humans are much more flexibly intelligent than other species. However, William James set this common-sense view on its head. He argued paradoxically that human behavior is more flexibly intelligent than that of other animals because humans have *more* instincts than other animals have, not fewer. Why should more instincts make humans more intelligent? James suggested that each module is a circuit with a distinct problem-solving ability tailored to particular piece of the world and relevant to a type of problem. The more such circuits one can link up, the broader the range of problems that can be solved. Humans tend to be blind to the existence of these instincts, however, precisely because they work so well — because they process information so effortlessly, automatically, and nonconsciously. They structure our thought so powerfully, James argued, that it can be difficult to imagine how things could be otherwise.

Humans take normal behavior so much for granted that they feel impelled to explain only *abnormal* behavior. For example, we usually do not ask why people breathe, eat, or are attracted to beautiful sexual partners — these responses are natural — instead, we ask why someone holds the breath, refuses to eat, or seems unmoved by the prospect of sex. As humans, we believe that normal behavior needs no explanation; but as scientists, explanation is in fact the goal: to describe all the programs or circuits in the human mind that cause humans to do all the normal things they do.

Blindness to our own instincts makes the study of psychology difficult because it makes the central object of study almost invisible. To get past this problem and to awaken ourselves to the real scientific task that confronts psychologists, James (1890) suggested that one try to make the "natural seem strange":

> It takes . . . a mind debauched by learning to carry the process of making the natural seem strange, so far as to ask for the *why* of any instinctive human act. To the metaphysician alone can such questions occur as: Why do we smile, when pleased, and not scowl? Why are we unable to talk to a crowd as we talk to a single friend? Why does a particular maiden turn our wits so upside-down? The common man can only say, *Of course* we smile, *of course* our heart palpitates at the sight of the crowd, *of course* we love the maiden, that beautiful soul clad in that perfect form, so palpably and flagrantly made for all eternity to be loved!
>
> And so, probably, does each animal feel about the particular things it tends to do in the presence of particular objects. . . . To the lion it is the lioness which is made to be loved; to the bear, the she-bear. To the broody hen the notion would probably seem monstrous that there should be a creature in the world to whom a nestful of eggs was not the utterly fascinating and precious and never-to-be-too-much-sat-upon object which it is to her.
>
> Thus we may be sure, that, however mysterious some animals' instincts may appear to us, our instincts will appear no less mysterious to them.

William James' views were a century ahead of their time; psychologists are only now beginning to explore the immensely intricate architecture of the human mind and to decode its programs. As James suggested, it took the emergence of the scientific study of animal minds, brains, and behavior (variously called **ethology**, animal behavior, behavioral ecology, or **sociobiology**) to propel the study of humans. Studying other species' contrasting competences and behaviors awakened researchers to a huge range of natural human competences and distinctive human behavior that had previously been ignored. Making the natural seem strange is unnatural, yet it is a pivotal part of the enterprise.

Until recently, most psychologists avoided the study of natural competences. Social psychologists, for example, were primarily interested in finding phenomena "that would surprise their grandmothers." Many cognitive psychologists spend more time studying how humans solve problems they are poor at (such as learning math or playing chess) than problems they are good at (for example, abilities to see, speak, regard someone as beautiful, reciprocate a favor, fear disease, fall in love, initiate an attack, experience moral outrage, navigate a landscape). Our natural competences are possible only because there is a vast and heterogeneous array of complex computational machinery supporting and regulating these activities. Just as modern personal computers come equipped with a variety of distinct programs to perform diverse tasks — a word processor, a spreadsheet, an address database, and so on — humans come equipped with a variety of task-specialized mental programs that switch off and on in different situations and cause us to fall in love, feel hungry, resent being cheated, deduce the meaning of a new word, and so on. The machinery works so well that humans do not realize that it exists; thus, we suffer from "instinct blindness." Psychologists are only now beginning to study some of the most interesting machinery in the human mind.

An evolutionary approach provides powerful lenses that correct for instinct blindness. It allows researchers to recognize what natural competences humans are likely to be equipped with; it indicates that the human mind is likely to contain a far vaster collection of these instincts, circuits, or competences than anyone even a decade ago suspected; and, most important, it provides specific and detailed theories of their designs. Einstein once commented "It is the theory which decides what we can observe." Evolutionary theory is valuable for psychologists who are studying a biological system of fantastic complexity because it allows researchers to observe evolved mental programs they otherwise would not have thought to look for, and so makes the intricate outlines of the mind's design stand out in sharp relief from the sea of incidental properties. Theories of adaptive problems can guide the search for the cognitive programs that solve them; knowing what cognitive programs exist can, in turn, guide the search for their neural basis (Figure 4.1).

THE STANDARD SOCIAL SCIENCE MODEL

Don Symons, one of the pioneers of evolutionary psychology, is fond of saying that you cannot understand what a person is saying unless you understand who that person is arguing with. Applying evolutionary biology to the study of the mind has brought most evolutionary psychologists into conflict with a traditional view of its structure, which arose long before Darwin. This view is no historical relic: It remains highly influential in psychology, anthropology, sociology, and the wider culture more than a century after Darwin and William James wrote. To understand evolutionary psychology, it is important to understand the view that evolutionary psychology is displacing; this view can be termed the **Standard Social Science Model** (Tooby and Cosmides, 1992).

Both before and after Darwin, a common view among philosophers and scientists has been that the human mind resembles a blank slate, virtually free of content until written on by the hand of experience. According to Aquinas, there is "nothing in the intellect which was not previously in the senses." Working within this framework, the philosophers known as the **British Empiricists** and their successors produced elaborate theories about how experience, refracted through a small handful of innate mental procedures, inscribed content onto the mental slate. David Hume's view was typical, and set the pattern for many later psychological and social science theories: ". . . there appear to be only three principles of connexion among ideas, namely *Resemblance*, *Contiguity* in time or place, and *Cause or Effect*."

Over the years, the technological metaphor used to describe the structure of the human mind has been consistently updated from blank slate to switchboard to general purpose computer to connectionist net, but the central tenet of the Empiricist views has remained the same. Indeed, it remains the reigning orthodoxy in most areas of psychology and in the social sciences. According to this orthodoxy, all the specific content of the human mind originally derives from the outside — from the environment and the social world — and the evolved architecture of the mind consists solely or predominantly of a small number of general-purpose mechanisms that are content independent; researchers refer to the mechanisms with terms such as learning, induction, intelligence, imitation, rationality, the capacity for culture, socialization, or simply culture.

According to this view, the same mechanisms are thought to govern how one acquires a language, how one learns to recognize emotional expressions, how one thinks about incest, or how one acquires ideas and attitudes about friends and reciprocity; indeed, they govern everything but perception, which is the conduit by which experience pours into the mind. The mechanisms that govern reasoning, learning, and

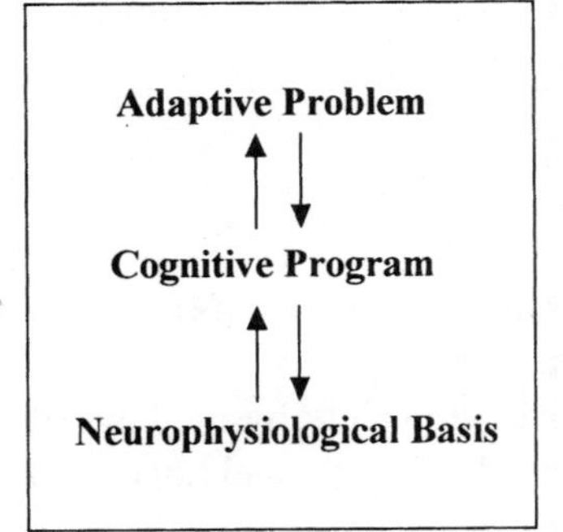

FIGURE 4.1

Three complementary levels of explanation in evolutionary psychology. Inferences (*arrows*) can be made from one level to another.

memory are assumed to operate uniformly, according to unchanging principles, regardless of the content they are operating on or the larger category or domain (that is, the topic) involved. For this reason, they are described as *content-independent* or **domain-general**. Such mechanisms, by definition, have no preexisting content built into their procedures, are not designed to construct certain mental contents more readily than others, and have no features specialized for processing particular kinds of content. Because these hypothetical mental mechanisms have no content to impart, it follows that all the particulars of what we think and feel are derived externally from the physical and social world. The social world organizes and injects meaning into individual minds, but the universal human psychological architecture has no distinctive structure that organizes the social world or imbues it with characteristic meanings. According to the Standard Social Science Model, the contents of human minds are primarily (or entirely) *free social constructions*, and the social sciences and human behavior are autonomous and disconnected from any evolutionary or biological foundation (Tooby and Cosmides, 1992). Humans are held to be the products of culture, not of instinct. Human nature is only the capacity to absorb culture — it has no other character.

Three decades of progress and convergence in cognitive psychology, evolutionary biology, and neuroscience have shown that this view of the human mind is radically defective. Evolutionary psychology provides an alternative framework that is beginning to replace the standard model. In the new view, all normal human minds reliably develop a standard collection of reasoning and regulatory circuits that are functionally specialized and, frequently, are designed to operate specifically within a particular domain (for example, sexual behavior, foods, navigation). That is, they are often **domain-specific**. The circuits organize the way humans interpret experiences, inject certain recurrent concepts and motivations into mental life, and provide universal frames of meaning that allow understanding of the actions and intentions of others. According to this view, beneath the level of surface variability, all humans share certain views and assumptions about the nature of the world and human action by virtue of evolved, universal cognitive programs. Concepts such as jealousy, friendship, beauty, and so on are not cultural inventions but, rather, cultural elaborations of universal features of the human mind.

BACK TO BASICS — FIVE BIOLOGICAL PRINCIPLES

How did evolutionary psychologists arrive at this view? When rethinking a field, it is sometimes necessary to go back to first principles, to ask basic questions: What is behavior? What do we mean by "mind"? How can something as intangible as a mind have evolved? What is its relation to the brain? The answers to such questions provide the framework within which evolutionary psychologists operate.

Psychology is the branch of biology that studies (1) brains, (2) how brains process information, and (3) how the brain's information-processing programs generate behavior. Psychology is a branch of biology because the human brain is a biological structure, a product of evolution. Once one realizes that psychology is a branch of biology, inferential tools developed in biology — its theories, principles, and observations — can be used to understand psychology. There are five basic principles, all drawn from biology, that evolutionary psychologists apply in their attempts to understand the design of the human mind. The five principles can be applied to *any* topic in psychology.

The principles organize observations in a way that allows one to see connections between areas as seemingly diverse as vision, reasoning, and sexuality.

Principle 1

The brain is a physical system functioning as a computer and has circuits designed to generate behavior appropriate to environmental circumstances.

The brain is a physical system whose operation is governed solely by the laws of chemistry and physics. What does this mean? It means that all thoughts and hopes and dreams and feelings are produced by chemical reactions going on in one's head (a sobering thought). Moreover, the brain's function is to process information. In other words, it is a computer that is made of organic (carbon-based) compounds rather than silicon chips. The brain is made of cells: primarily **neurons** and their supporting structures. Neurons are cells that were modified over the course of evolution so that they are specialized for the transmission of information. Electrochemical reactions cause neurons to fire.

Neurons are connected to one another in a highly organized way. One can think of these connections as circuits — just as a computer has circuits. The circuits determine how the brain processes information, just as the circuits in a computer determine how it processes information. Neural circuits in the brain are connected to sets of neurons that run throughout the body. Some neurons are connected to sensory receptors, such as the retinas of one's eyes. Others are connected to muscles. **Sensory receptors** are cells that are specialized for gathering information from the outer world and from other parts of the body; for example, you can feel your stomach churn because the stomach has sensory receptors on it, but you cannot feel your spleen, because it lacks receptors. Sensory receptors are connected to neurons that transmit the information to your brain. Other neurons send information from your brain to motor neurons. **Motor neurons** are connected to muscles; they cause muscles to move. This movement is called *behavior*.

Organisms that do not move do not have brains — trees do not have brains, bushes do not have brains, grasses do not have brains. In fact, some animals do not move during certain stages of their lives, and during those stages, *they* do not have brains. The sea squirt, for example, is a marine invertebrate (backbone-lacking) animal. During the early stage of its life cycle, the sea squirt swims around looking for a good place to attach itself permanently. Once it finds the right rock and attaches itself to it, it does not need its brain because it will never need to move again. So it eats (resorbs) most of its brain! (After all, why waste energy on a now useless organ? Better to get a good meal out of it.)

In short, the circuits of the brain are designed to generate motion — behavior — in response to information from the environment. The function of the brain, the "wet computer," is to generate behavior that is appropriate to your environmental circumstances.

Principle 2

Neural circuits were designed by natural selection to solve problems that ancestors faced during the species' evolutionary history.

To say that the function of the brain is to generate behavior that is appropriate to your environmental circumstances is not saying much, unless one has a definition of what appropriate means. What counts as appropriate behavior?

Appropriate has different meanings for different organisms. Humans have sensory receptors that are stimulated by the sight and smell of feces — to put it more bluntly, we can see and smell dung. So can a dung fly. But on detecting the presence of feces in the environment, a human's appropriate behavior differs from what is appropriate for the dung fly. On smelling feces, appropriate behavior for a female dung fly is to move toward the feces, land on them, and lay her eggs. Feces are food for a dung fly larva — therefore, appropriate behavior for a dung fly larva is to eat dung. And, because female dung flies hang out near piles of dung, appropriate behavior for a male dung fly is to buzz around the piles and try to mate; for a male dung fly, a pile of dung is a sexually exciting haunt, much as a disco or a bar might be for humans who live in Western societies.

But for human ancestors, feces were (and are) a source of contagious diseases. For humans, feces are not food. They are not a good place to raise children, nor are they a good place to look for a sexual partner. The appropriate behavior for a human is to move away from the source of the smell. Humans may also form the cross-cultural, universal *disgust expression* with facial muscles — the nose wrinkles to protect the eyes and nose from the volatiles, and the tongue protrudes slightly as if one were ejecting something from the mouth.

For humans, the pile of dung is disgusting. For a female dung fly, looking for a good neighborhood and a nice house for raising her children, the pile of dung is a beautiful vision — a mansion. (Seeing a pile of dung as a mansion — perhaps *that* is what William James meant by making the natural seem strange!)

The point is, environments do not, themselves, specify what counts as appropriate behavior. In other words, it is not a scientific explanation to say that "The environment made me do it!" In principle, a computer program or neural circuit could be designed to link *any* given stimulus in the environment to any kind of behavior. For feces, evolution has equipped two species with two different circuits that cause opposite responses for the same stimulus. So, the explanation for a behavior cannot be in the stimulus alone, but must also be in the nature of the circuits each species has. Which behavior a stimulus gives rise to is a function of the neural circuitry of the organism, which means that a designer of brains could have made a person who licks her chops and sets the table when she smells a nice fresh pile of dung.

But what did the actual, natural designer of the human brain do, and why? Why do humans find fruit sweet and dung disgusting? How did humans get the circuits they have, rather than the circuits the dung fly has?

In talking about a home computer, one can answer the question simply: The circuits were designed by an engineer, and the engineer designed them one way rather than another so they would solve problems that the engineer *wanted* them to solve — problems such as adding or subtracting or accessing a particular address in the computer's memory. Human neural circuits were also designed to solve problems, but they were not designed by an engineer. They were designed by the *evolutionary process* to solve problems that were important from an evolutionary standpoint during our evolutionary history.

Toward what good or end were evolved mental programs designed by natural selection to work? For one thing, natural selection does *not* work for the good of the species or for the perpetuation of the species, as many people think. *Natural selection* is a process in which a mutant or new design feature *causes its own spread through a population over multiple generations* (which can happen regardless of what such a spread does to the welfare or even the long-term survival of the species). To understand evolution, one could think of natural selection as the "eat dung and die" principle. All animals need neural circuits that govern what they eat — knowing what is safe to eat is a problem that all animals must solve. For humans, feces are not safe to eat — they are a source of contagious diseases. An ancestral human who had mutant neural circuits that made dung smell appetizing would increase the probability of contracting a disease. If that human got sick as a result, he would be too tired to find much food, too exhausted to go looking for a mate, and might even die an untimely death. In contrast, a person with different neural circuits — circuits that made him avoid feces — would get sick less often than persons lacking such circuits. He would therefore have more time to find food and mates and would live a longer life. The first person would eat dung and die; the second would avoid it and live. As a result, the dung-eater would have fewer children than the dung-avoider. Since the neural circuitry of children tends to resemble that of their parents, there would be fewer dung-eaters and more dung-avoiders in the next generation. As this process continued, generation after generation, the dung-eaters would eventually disappear from the population. Why? They ate dung and died out! The only kind of people left in the population would be the ones descended from the dung-avoiders (that is, persons with very complex and sophisticated disgust and food choice circuitry). No one who has neural circuits that make dung delicious would be left.

The reason humans have one set of circuits rather than another is that the circuits humans have were better at causing their own spread over many ancestral generations than were alternative circuits implementing alternative behavioral patterns. The inherited tendency to cause design features, such as neural circuits, to increase their own frequency over generations is the "good" (the "engineering goal") that governs which design natural selection builds into a species. The route by which design features are built is through increasing an individual's reproduction or the reproduction of family members — individuals who might be carrying the same design features. Adaptive problems are situations that interfere with an individual's reproduction or survival, or the survival and reproduction of an individual's relatives. The brain is a naturally constructed computational system whose function is to solve adaptive information-processing problems (such as face recognition, threat interpretation, language acquisition, navigation). Over evolutionary time, the brain's circuits were cumulatively added because they reasoned or processed information in a way that enhanced the adaptive regulation of behavior and physiology.

However, human circuits were not designed to solve *any* kind of problem. They were designed to solve *adaptive* problems. **Adaptive problems** have two defining characteristics. First, they are the problems that cropped up again and again during the evolutionary history of a species. Second, they are problems whose solution enhanced or facilitated the *reproduction* of individual organisms — however indirect the causal chain may be and however small the effect on number of offspring produced, because differential reproduction (and not survival itself) is the engine that drives natural selection.

Consider the fate of a circuit that had the effect, on average, of enhancing the reproductive rate of the organisms that sported it, but in so doing shortened the average lifespan of the organisms, for example, a circuit that causes mothers to risk death to save their children. If this effect were heritable, its frequency in the population would increase. In contrast, any circuit that had the effect of decreasing the reproductive rate of the organisms that had that circuit would eventually disappear from the population. Most adaptive problems have to do with how an organism makes its living or avoids difficulties: what it eats, what eats it, who it mates with, who it socializes with, how it communicates, how it avoids being attacked, and so on. The *only* kind of problems that natural selection can design circuits for solving are adaptive problems.

Obviously, humans now are able to solve problems that no hunter-gatherer ever had to solve — learn higher mathematics, drive cars, and use computers. Our ability to solve other kinds of problems is a side-effect or by-product of circuits that were designed to solve adaptive problems. For example, when our ancestors became bipedal — when they started walking on two legs instead of four — they had to develop a very good sense of balance. Indeed, we have in the inner ear very intricate mechanisms that allow us to achieve an excellent sense of balance. We can balance well on two legs while moving, which means that we can do other things besides walk — skateboard or ride the waves on a surfboard. Our hunter-gatherer ancestors were not tunneling through curls along the shores of what would become Kenya. That we can surf and skateboard is merely a by-product of adaptations designed for balancing while walking on two legs.

Principle 3

Most of what goes on in the mind is hidden; thus, most problems that seem easy to solve are actually very difficult to solve — they require very complicated circuitry.

Humans are not and cannot become consciously aware of most of the brain's ongoing activities. To illustrate by analogy, think of your brain as the entire federal government and as the *self* that is consciously experienced as *self* as the President of the United States. If you were the President, how would you know what is going on in the world? Members of the Cabinet, such as the Secretary of Defense, would tell you things — for example, that the Bosnian Serbs are violating their cease-fire agreement. How do members of the Cabinet know these things? Thousands of bureaucrats in the State Department, thousands of Central Intelligence Agency operatives in Serbia and other parts of the world, thousands of troops stationed overseas, and hundreds of investigative reporters are gathering and evaluating enormous amounts of information from all over the world. But you, as the President, do not — and in fact, cannot — know what each of the thousands of individuals were doing while each gathered the information during the course of a few months — what each of them saw, what each of them read, to whom each of them talked, what conversations were clandestinely taped, what offices were bugged. All you, as the President, know is the final conclusion that the Secretary of Defense has on the basis of the information that was passed to him. And all he knows is what other high-level officials passed to him. And so on. In fact, no single individual knows *all* the facts of the situation because the facts are distributed among thousands of people. Moreover, each of the thousands of individuals involved knows all kinds of details about the situation that were decided not important enough to pass to higher levels.

So it is with conscious experience. The only things a person becomes aware of are a few high-level conclusions passed on by thousands and thousands of specialized mechanisms: some that are gathering sensory information from the world, and others that are analyzing and evaluating the information, checking for inconsistencies, filling in the blanks, figuring out what it means.

Any scientist who is studying the human mind must keep this fact in mind. In figuring out how the mind works, one's conscious experience of self and of the world can suggest some valuable hypotheses; but the same intuitions can also be seriously misleading. They can fool a person into thinking that neural circuitry is simpler than it really is. For example, conscious experience tells you that seeing is simple: You open your eyes, light hits your retinas, and — voila! — you see. The process is effortless, automatic, reliable, fast, unconscious, and requires no explicit instruction. But the apparent simplicity is deceptive. Retinas are two-dimensional sheets of light-sensitive cells covering the inside back of the eyeballs. Figuring out only on the basis of the light-dependent chemical reactions occurring in the flat arrays of cells what three-dimensional objects exist in the world poses enormously complex problems — so complex, in fact, that no computer programmer has yet been able to create a robot that can see even remotely as well as humans do routinely. One sees with the brain, not just the eyes, and the brain contains a vast array of dedicated, special-purpose circuits, each set specialized for solving a different component of the problem. You need all kinds of circuits simply to see your mother walk, for example. You have circuits that are specialized for (1) analyzing the *shape* of objects; (2) perceiving the presence of *motion*; (3) detecting the *direction* of motion; (4) judging *distance*; (5) analyzing *color*; (6) recognizing an object as *human*; and (7) *identifying* the face you see as Mom's face, rather than someone else's. Each individual circuit is shouting its information to higher-level circuits, which check the facts generated by one circuit against the facts generated by the others, and resolve contradictions. The conclusions are then handed over to still higher-level circuits that piece the facts together and hand the final report to the President — your consciousness. But all this "president" ever becomes aware of is the sight of *Mom walking*. Although each circuit is specialized for solving a delimited task, the circuits work together to produce a coordinated functional outcome — in this case, your conscious experience of the visual world. Seeing is effortless, automatic, reliable, and fast precisely because humans embody this complicated, dedicated machinery.

Our intuitions can deceive us. Conscious experience of an activity as easy or natural can lead people to grossly underestimate the complexity of the circuits that make the experience possible. Doing what comes naturally, effortlessly, or automatically is rarely simple from an engineering point of view. To find someone beautiful, to fall in love, to feel jealous — all these events can seem as simple and automatic and effortless as opening our eyes and seeing, so simple that it seems like there is nothing to explain about such a natural response. But these activities feel effortless only because there is a vast array of complex neural circuitry supporting and regulating them. Other species face other problems, and so their minds compute other things: spider webs, bat echolocation, and so on.

Principle 4

Different neural circuits are specialized for solving different adaptive problems.

A basic engineering principle is that the same machine is rarely capable of solving two different problems equally well. Screwdrivers and saws, for instance, both exist because each solves a particular problem better than the other. Just imagine trying to cut planks of wood with a screwdriver or to turn screws with a saw.

Our body is divided into organs, such as the heart and the liver, for exactly this principle. Pumping blood throughout the body and detoxifying poisons are two very different problems. Consequently, the body has a different machine for solving each problem. The design of the heart is specialized for pumping blood; the design of the liver is specialized for detoxifying poisons. The liver cannot function as a pump, and the heart is not good at detoxifying poisons.

For the same reason, the mind consists of a large number of circuits that are *functionally specialized*. Some neural circuits are designed specialized for vision. Other neural circuits are specialized for hearing — all they do is detect changes in air pressure and extract information from it. They do not participate in vision, vomiting, vanity, vengeance, or anything else. Still other neural circuits are specialized for sexual attraction — that is, they govern what you find sexually arousing, what you regard as beautiful, who you'd like to date, and so on.

Humans have all these specialized neural circuits because the same mechanism is rarely capable of solving different adaptive problems. For example, all humans have neural circuitry designed to choose nutritious food on the basis of taste and smell — circuitry that governs our food choice. But imagine a woman who used the same neural circuitry to choose a mate. She would choose a strange mate indeed! (Perhaps a huge chocolate bar?) To solve the adaptive problem of finding the right mate, our choices must be guided by *qualitatively different standards* than those used to choose the right food or the right habitat. Consequently, the brain must be composed of a large collection of circuits, with different circuits specialized for solving different problems. One can think of each specialized circuit as a minicomputer that is dedicated to solving one problem. Such dedicated minicomputers are sometimes called *modules*. There is, then, a sense in which one can view the brain as a collection of dedicated minicomputers — a collection of modules. Of course, there must be circuits whose design is specialized for integrating the output of the dedicated minicomputers to produce behavior. So, more precisely, one can view the brain as a collection of dedicated minicomputers whose operations are *functionally integrated* to produce behavior.

Psychologists have long known that the human mind contains circuits that are specialized for different modes of perception, but until recently it was thought that perception and, perhaps, language were the only activities caused by specialized cognitive processes (Fodor, 1983). Other cognitive functions — learning, reasoning, decision-making — were thought to be accomplished by general-purpose circuits. Prime candidates were rational algorithms: ones that implement formal methods for inductive and deductive reasoning, such as Bayes' rule or the propositional calculus (a formal logic). **General intelligence** — a hypothetical faculty composed of a few simple reasoning circuits that are content-independent and for general purposes — was thought to be the engine that generates solutions to reasoning problems. The flexibility of human reasoning (that is, one's ability to solve many different kinds of problems) was thought to be evidence for the generality of the circuits that generate reasoning.

An evolutionary perspective suggests that human intelligence is not a product of a blank slate connected to content-independent mechanisms that operate according to rational algorithms (Tooby and Cosmides, 1992). Cognitive programs are designed to mesh with the features of the environments in which they evolved, and so they embody information about the stably recurring properties of these ancestral worlds. For example, the snake-phobia mechanism in humans embodies information about the abstract appearance of snakes and about the danger that snakes can pose. Similarly, human color-constancy mechanisms are calibrated to changes in natural — that is, Sun-caused — illumination; as a result, grass looks green at both high noon and sunset, even though the spectral properties of the light it reflects have changed dramatically. In contrast, **content-independent** (rational) problem-solving methods do not embody specific information about the structure of various domains of the world relevant to different problem types because content-independent methods, by definition, operate in the same manner regardless of what the situation is like or what the content of the problem is.

Figure 4.2 shows two rules of inference from the **propositional calculus**, a system that allows one to deduce true conclusions from true premises, no matter what the subject matter of the premises is — no matter what P and Q refer to. **Bayes' rule**, an equation

FIGURE 4.2

Two rules of inference of the propositional calculus. The rules are valid: Given true premises, they generate true conclusions. The rules are also content-independent: *P* and *Q* can stand for any proposition, no matter what it is about.

Modus ponens:	*an example:*
If P then Q	If you slept, then you had dreams
P	You slept
therefore Q	therefore you had dreams
Modus tollens:	
If P then Q	If you slept, then you had dreams
not-Q	You did not have dreams
therefore not-P	therefore you did not sleep

for computing the probability that a hypothesis is true given specific observations, is also content-independent. It can be applied indiscriminately to medical diagnosis, card games, hunting success, or any other subject. It contains no knowledge that is *specific* to the activity considered, so it cannot cause correct judgments that would apply to mate choice, for example, but not to hunting (that is the price of content-independence).

Evolved problem-solvers, in contrast, are equipped with "crib sheets": They come to a problem already knowing a lot about it. For example, a newborn's brain has response systems that expect faces to be present in the environment: Babies less than 10 minutes old turn their eyes and head in response to face-like patterns but not to scrambled versions of the same pattern with identical spatial frequencies (Johnson and Morton, 1991). Infants' brains make strong assumptions about how the world works and what kinds of things it contains, even at 2.5 months (the point at which infants can see well enough to be tested). The infant's brain assumes, for example, that the world contains rigid objects that are continuous in space and time, and the brain has preferred ways of dividing the world into separate objects (Baillergeon, 1986; Spelke, 1990). Ignoring shape, color, and texture, the infant's brain treats any surface that is cohesive, bounded, and moves as a unit as a single object. When one solid object appears to pass through another, the infants are surprised, just as an adult would be. A mind that was a blank slate — a truly "open-minded" system — would be undisturbed by such displays.

In watching objects interact, babies less than 1 year old distinguish causal events (objects causing others to move by bumping into them) from noncausal ones that have similar spatiotemporal properties (that is, objects that start and stop moving in sequence but do not touch each other); they distinguish objects that move only when acted on from objects that are capable of self-generated motion (the so-called inanimate/animate distinction); they assume that the self-propelled movement of animate objects is caused by invisible internal states — goals and intentions — whose presence must be inferred because internal states cannot be seen (Baron-Cohen, 1995; Leslie, 1988; 1994). Toddlers, like adults, have a well-developed "mind-reading" system that uses eye direction and movement to infer what other people want, know, and believe

(Baron-Cohen, 1995). (When this system is impaired, as in autism, the child cannot understand what others think.) When an adult utters a word-like sound while pointing to an object, toddlers assume the word refers to the whole object, not one of its parts (Markman, 1989).

If the evolved human mind lacked these privileged hypotheses about faces, objects, physical causality, other minds, word meanings, and so on, a developing child could learn very little about its environment. For example, an autistic child who has a normal IQ and intact perceptual systems is unable to make simple inferences about mental states (Baron-Cohen, 1995); children with Williams syndrome are profoundly retarded and have difficulty learning even very simple spatial tasks, but are good at inferring other people's mental states and complex language skills — some of their specialized reasoning mechanisms are damaged, but their mind-reading system is intact.

Different problems require different crib sheets. For example, knowledge about intentions, beliefs, and desires, which allows one to infer the behavior of persons, is misleading if applied to inanimate objects. Two cognitive programs with different crib sheets perform better than one when the crib sheet that helps solve problems in one domain is misleading in another area. Human minds evolved to understand inanimate objects differently than animate objects — entities whose behavior is caused by invisible internal states such as desires and beliefs, which suggests that many evolved computational mechanisms are domain-specific — activated in some domains but not others. Some mechanisms embody rational, content-independent methods, but others have special-purpose, inferential procedures that respond to content types — procedures that work well within the stable ecological structure of a particular domain even though they might lead to false or contradictory inferences if they were activated outside that domain. Activating programs designed to interpret behavior in terms of beliefs and desires when one is thinking about inanimate objects leads to confusion and error.

The more distinct crib sheets a system has, the more problems the system can solve. A brain equipped with a multiplicity of specialized inference engines can generate sophisticated behavior that is sensitively tuned to the varieties of problems generated by its environment. In this view, the flexibility and power often attributed to content-independent problem-solving methods is illusory. If all else is equal, a content-rich system can infer more than a content-poor one can.

Machines limited to executing rational procedures such as Bayes' rule, modus ponens (see Figure 4.2), and others derived from mathematics or logic are computationally weak compared to the system outlined above (Tooby and Cosmides, 1992). The theories of rationality these mechanisms embody are environment-free, that is, designed to produce valid inferences in *all* domains, regardless of the nature of the local world the organism lives in or the type of problem the organism is facing. The theories can be applied to a wide variety of domains because they lack any information that would be helpful in one domain but not in another. Having no crib sheets, no domain-specific information, there is little they can deduce about a domain; having no privileged hypotheses, there is little they can induce before their operation is hijacked by the necessity of considering an endless string of possibilities. The difference between *domain-specific* methods and *domain-independent* ones is akin to the difference between experts and novices: Experts can solve problems faster and more efficiently than novices because they already know a lot about the problem domain.

William James' view of the mind, ignored for much of the twentieth century, is being vindicated today. There is now evidence for the existence of circuits that are specialized for reasoning about objects, physical causality, number, the biological world, the beliefs and motivations of other individuals, and social interactions (Hirschfeld and Gelman, 1994). For example, it is now known that the learning circuits that govern the acquisition of language are different from circuits that govern the acquisition of food aversions, and both are different from the learning circuits that govern the acquisition of snake phobias (Garcia, 1990; Pinker, 1994; Mineka and Cooke, 1985; Ohman et al., 1985). Current literature is full of examples of how each species is governed by a multitude of distinct modules, instincts, or cognitive programs that guide learning differently for different topics.

Instinct is often thought of as the polar opposite of reasoning and learning. *Homo sapiens* is thought of as the rational animal — a species whose instincts were erased by evolution because the ability to reason and learn rendered them useless or harmful. But human reasoning circuits and learning circuits have the following five properties: (1) they are complexly structured for solving a specific type of adaptive problem; (2) they reliably develop in all normal human beings; (3) they develop without any conscious effort and in the absence of any formal instruction; (4) they are applied without any conscious awareness of their underlying logic; and (5) they are distinct from more general abilities to process information or behave intelligently. In other words, all have the hallmarks of what one usually thinks of as an instinct (Pinker, 1994). In fact, one can think of these special-purpose computational systems as *reasoning instincts* and *learning instincts*. These systems make certain kinds of inferences just as easy, effortless, and natural to humans as spinning a web to a spider or dead-reckoning to a desert ant (see Chapter 1). Students and social scientists often ask whether a behavior was caused by instinct or learning. A better question is, What are the design features of the instinct that caused the learning?

Principle 5

Modern skulls house Stone Age minds.

Natural selection, the process that designed the human brain, takes a long time to design a circuit of any complexity. The time it takes to build circuits that are suited to a given environment is so slow it is hard to even imagine — it is like a stone being sculpted by wind-blown sand. Even relatively simple changes can take thousands of generations.

For this reason, the environment that humans — and, therefore, human *minds* — evolved in was very different from the modern environment. Human ancestors spent more than 99% of their evolutionary history living in **hunter-gatherer** societies — our forebears lived in small, nomadic bands of a few dozen individuals who got all their food each day by gathering plants or by hunting animals. Each of our ancestors was, in effect, on a camping trip that lasted an entire lifetime, and the world of our foraging ancestors endured for millions of years.

Generation after generation, for 2 (or more) million years, natural selection slowly sculpted the human brain, favoring circuitry that was good at solving the day-to-day problems of our hunter-gatherer ancestors — problems such as finding mates, hunting animals, gathering plant foods, negotiating with friends, defending against aggression, raising children, choosing a good habitat, and so on. Those whose circuits were better designed for solving these problems consistently left more children, and so we and our entire species are descended from them, not from others who lived during ancestral times but who lacked well-designed circuits.

Humans lived as hunter-gatherers *1000 times longer* than they lived in any other way. The world that seems so familiar to us, a world with roads, schools, grocery stores, factories, farms, and nation-states, has existed for only an eyeblink when compared to our entire evolutionary history. The computer age is only a little older than the typical college student, and the industrial revolution is a mere 200 years old. Agriculture first appeared in some regions of the globe only 10,000 years ago, and it was not until about 5000 years ago that as many as half the human population engaged in farming rather than hunting and gathering. Natural selection is a slow process; there simply have not been enough generations for it to select new circuits that are well adapted to postindustrial life.

In other words, our modern skulls house a Stone Age mind. The *key to understanding how the modern mind works* is to realize that its circuits were not designed to solve the day-to-day problems of a modern American; they were designed to solve the day-to-day problems of our hunter-gatherer ancestors. These Stone Age priorities produced a brain far better at solving some problems than others. For example, it is easier for us to deal with small, hunter-gatherer-band-sized groups of people than to deal with crowds of thousands; it is easier for us to learn to fear snakes than electric sockets, even though electric sockets pose a larger threat than snakes do in most American communities. In many cases, our brains are *better* at solving the kinds of problems our ancestors faced on the African savannas than they are at solving the more familiar tasks we face in a college classroom or a modern city. Saying that modern skulls house a Stone Age mind does not imply that our minds are unsophisticated. Quite the contrary — they are very sophisticated computers with circuits elegantly designed to solve the exotic kinds of problems our *ancestors* routinely faced.

Understanding that human modules, instincts, or cognitive programs evolved to solve the adaptive problems faced by our ancestors makes understanding the ancestral world important. Each cognitive program is an engineered device designed to solve problems by taking advantage of the stable structure of the problem environment. In the ancestral world, the so-called **environment of evolutionary adaptedness** (EEA), sweetness signaled nutrition, smooth skin indicated youth and health, objects moved in continuous smooth trajectories through three-dimensional space, and so on. Because mental adaptations or cognitive adaptations are designed to mesh with the relevant features of the EEA, understanding the engineering of the human mind requires understanding the structure of the EEA. Thus, the EEA is not a place or time as was the African savanna during the Pliocene or Pleistocene Epochs. The statistical composite of selection pressures — environmental properties — caused the design of an adaptation to be favored over alternative designs. The EEA for one adaptation, therefore, may be different from that for another. Conditions of terrestrial illumination, which form part of the EEA for the vertebrate eye, remained relatively constant for hundreds of millions of years (until the invention of the incandescent bulb). In contrast, the EEA that selected mechanisms that cause human males to provide food and other resources to their offspring — a situation that departs from the typical mammalian pattern — appears to be less than 2 million years old.

The five principles outlined here are tools for thinking about psychology and can be applied to any topic: sex and sexuality, how and why people cooperate, whether people are rational, how babies see the world, conformity, aggression, hearing, vision, sleeping, eating, hypnosis, schizophrenia, and on and on. The framework the principles provide links areas of study and saves one from drowning in irrelevant particulars. Whenever one tries to understand some aspect of human behavior, these principles encourage one to ask the following fundamental questions:

1. Where in the brain are the relevant circuits and how, physically, do they work?
2. What kind of information is being processed by the circuits?

3. What information-processing programs do the circuits embody?
4. What were the circuits designed to accomplish (in a hunter-gatherer context)?

UNDERSTANDING THE DESIGN OF ORGANISMS

Adaptationist Logic and Evolutionary Psychology

Phylogenetic Versus Adaptationist Explanations The goal of Darwin's theory was to explain phenotypic design: Why do the beaks of finches differ from one species to another? Why do animals expend energy attracting mates that could be spent on survival? Why are human facial expressions of emotion similar to the expressions of other primates?

Two of the most important evolutionary principles accounting for the characteristics of animals are (1) common descent, and (2) adaptation driven by natural selection. If all humans are related to one another and to all other species, by virtue of common descent, one might expect to find similarities between humans and their closest primate relatives. The **phylogenetic approach** has a long history in psychology — it prompts the search for phylogenetic continuities implied by the inheritance of homologous features from common ancestors.

An **adaptationist approach** to psychology leads to the search for adaptive design, which entails understanding how various design features constitute organic machinery designed by selection to solve a problem. Such a functional unit or mechanism is an adaptation. Although many adaptations are similar among closely related species (for example, eyes), adaptationism often entails the examination of niche-differentiated mental abilities unique to the species being investigated. Elephants have trunks, humans have language, bats have echolocation — none of these features are shared by other mammals. What makes the organized differences comprehensible is an adaptationist approach. A book by George Williams (1966), *Adaptation and Natural Selection*, clarified the logic of adaptationism. This work provided the tools that allowed the construction of modern evolutionary psychology. Evolutionary psychology can be thought of as the application of adaptationist logic to the study of the architecture of the human mind.

Why Does Structure Reflect Function? In evolutionary biology, several different levels of explanation are complementary and mutually compatible. Explanation at one level (for example, adaptive function) does not preclude or invalidate explanations at another level (for example, neural, cognitive, social, cultural, or economic). Evolutionary psychologists use theories of adaptive function to guide their investigations of cognitive and neural structures. Why is this possible? The evolutionary process has two components: chance and natural selection. Natural selection is the only component of the evolutionary process that can introduce complex *functional* organization into a species' phenotype (Dawkins, 1986; Williams, 1966).

The function of the brain is to generate behavior that is sensitively contingent on information from an organism's environment. The brain, therefore, is an information-processing device. Neuroscientists study the physical structure of such devices, and cognitive psychologists study the information-processing programs realized by that structure. There is, however, another level of explanation — a functional level. In evolved systems, form follows function. The physical structure of the brain is there because it embodies a set of programs; a program is there because its set of procedures causes the solution to a problem that the species' ancestors commonly encountered. The functional level of explanation is essential for understanding how natural selection designs organisms.

An organism's phenotypic structure can be thought of as a collection of design features — micromachines, such as the functional components of the eye or liver. Over evolutionary time, new design features are added or discarded from the species' design because of their consequences. A design feature causes its own spread across generations if it has the consequence of solving adaptive problems: for instance, cross-generationally recurrent problems, such as detecting predators or detoxifying poisons, whose solution promotes reproduction. As another example, if more sensitive retinas appeared in one or a few individuals by chance mutation and allowed predators to be detected more quickly, individuals who had the more sensitive retinas would produce offspring at a higher rate than would individuals who lacked the retinas. By promoting the reproduction of its bearers, the more sensitive retina thereby *promotes its own spread across generations*, until it eventually replaces the earlier-model retina and becomes a universal feature of that species' design.

Hence, natural selection is a feedback process that chooses among alternative designs on the basis of *how well the designs function*. It is a hill-climbing process in which a design feature that solves an adaptive problem well can be outcompeted by a new design feature that solves the problem better. This process has produced exquisitely engineered biological machines — the vertebrate eye, photosynthetic pigments, efficient foraging algorithms, color-constancy systems — the performance of which are unrivaled by any machines yet designed by humans.

By selecting designs on the basis of how well they solve adaptive problems, the process engineers a tight fit between the function of a device and its structure. To understand this causal relationship, biologists had to develop a theoretical vocabulary that distinguishes between structure and function. In evolutionary biology, explanations that appeal to the structure of a device are sometimes called *proximate explanations*. When applied to psychology, these explanations include those that focus on genetic, biochemical, physiological, developmental, cognitive, social, and all other immediate causes of behavior. Explanations that appeal to the adaptive function of a device are sometimes called *distal* or *ultimate explanations* because they refer to causes that operated over evolutionary time.

Knowledge of Adaptive Function Is Necessary for Carving Nature at the Joints An organism's phenotype can be partitioned into *adaptations* (present because they were chosen), *by-products* (present because they are causally coupled to traits that were selected — for example, the whiteness of bone), and *noise* (injected by the stochastic components of evolution). Like other machines, only narrowly defined aspects of organisms fit together into functional systems — most ways of describing the system do not capture its functional properties. Unfortunately, some have misrepresented the well-supported claim that selection creates functional organization as the obviously false claim that all traits of organisms are functional — something no sensible evolutionary biologist would ever maintain.

Furthermore, not all behavior engaged in by organisms is adaptive. A taste for sweet things may have been adaptive in ancestral environments in which nutrient-rich fruit was scarce, but it can generate maladaptive behavior in a modern environment flush with fast-food restaurants. Moreover, once an information-processing mechanism exists, it can be deployed in activities that are unrelated to its original function. For example, because we have evolved learning mechanisms that cause language acquisition, we can learn to write, which is the pairing of already learned words with arbitrary visual symbols. But these learning mechanisms were not chosen *because* they caused writing.

Design Evidence Adaptations are problem-solving machines, and they can be identified using the same standards of evidence that one would use to recognize a human-made machine: design evidence. One can identify a machine as a television set, not a stove, by finding evidence of complex functional design for that purpose — showing, for example, that the machine has many coordinated design features (for example, antennas, cathode ray tubes) that are complexly specialized for transducing television waves and transforming them into a color bit map (a configuration that is unlikely to have risen solely by chance), whereas the machine has virtually no design features that would make it good at cooking food.

Complex functional design is also the hallmark of living adaptive machines. One can identify an aspect of the phenotype as an adaptation by showing that (1) it has many design features that are complexly specialized for solving an adaptive problem; (2) the phenotypic properties are unlikely to have arisen by chance alone; and (3) they are not better explained as the by-product of mechanisms designed to solve some alternative adaptive problem. Finding that an architectural element solves an adaptive problem with reliability, efficiency, and economy is prima facie evidence that one has located an adaptation (Williams, 1966).

Design evidence is important not only for explaining why a known mechanism exists but also for discovering new mechanisms that no one had previously thought to look for. Theories of problems that our ancestors had to be able to solve — theories of adaptive function — play an important role in guiding the search for unknown mental characteristics and machinery.

Researchers who study species from an adaptationist perspective adopt the stance of an engineer. For example, in discussing sonar in bats, Dawkins (1986, pp. 21–22) says: "I shall begin by posing a problem that the living machine faces, then I shall consider possible solutions to the problem that a sensible engineer might consider; I shall finally come to the solution that nature has actually adopted." Engineers figure out what problems they want to solve and then design machines that are capable of solving the problems in an efficient manner. Evolutionary biologists figure out what adaptive problems a given species encountered during its evolutionary history and then ask, What would a machine capable of solving these problems well under ancestral conditions look like? Against this background, evolutionary biologists empirically explore the design features of the evolved machines that comprise an organism. Definitions of adaptive problems do not, of course, uniquely specify the design of the mechanisms that solve the problems.

Because there are often many ways of achieving any solution, empirical studies are needed to decide which mechanisms nature has actually adopted. The more precisely one can define an adaptive information-processing problem — the "goal" of processing — the more clearly one can see what a mechanism capable of producing that solution would have to look like. This research strategy has dominated the study of vision, for example, so that it is now commonplace to think of the visual system as a collection of functionally integrated computational devices, each specialized for solving a different problem in scene analysis, for judging depth, detecting motion, analyzing shape from shading, and so on. In our own research (discussed below), we have applied this strategy to the study of social reasoning.

Nature and Nurture — An Adaptationist Perspective

To fully understand the concept of design evidence, one must consider how an adaptationist thinks about nature and nurture. Debates about the relative contribution during development of nature and nurture have been among the most contentious in psychology. The premises that underlie these debates are flawed, yet they are so deeply entrenched that many people have difficulty seeing that there are other ways to think about these issues.

Evolutionary psychology is *not* just another swing of the nature-nurture pendulum. A defining characteristic of the field is the explicit rejection of the usual nature-nurture dichotomies — instinct versus reasoning, innate versus learned, biological versus cultural. What effect the environment has on an organism depends critically on the details of the evolved cognitive architecture of the organism. For this reason, coherent "environmentalist" theories of human behavior all, necessarily, make nativist claims about the exact form of evolved human psychological mechanisms. For evolutionary psychologists, the real scientific issues concern the design, nature, and number of the evolved mechanisms, not biology versus culture, genes versus learning, or other malformed oppositions.

Several different nature-nurture issues are usually conflated. It is useful to examine them separately because some are nonissues and others are real issues.

Focus on Architecture At a certain level of abstraction, every species has a universal, species-typical evolved architecture. For example, one can open any page of the medical textbook, *Gray's Anatomy*, and find the design of evolved architecture described down to the minutest detail — not only do all humans have a heart, two lungs, a stomach, intestines, and so on, but the book describes human anatomy to the particulars of specific nerve connections. This is not to say there is no biochemical individuality. No two stomachs are exactly alike — they vary in quantitative properties such as size, shape, and how much hydrochloric acid they produce. But all humans have stomachs, and all human stomachs have the same basic *functional* design — each is attached at one end to an esophagus and at the other to the small intestine; each secretes the same chemicals necessary for digestion; and so on. Presumably, the same is true of the brain and, hence, of the evolved architecture of our cognitive programs — of the information-processing mechanisms that generate behavior. Evolutionary psychology seeks to characterize the universal, species-typical architecture of these mechanisms.

Cognitive architecture, like all aspects of the phenotype from molars to memory circuits, is the joint product of genes and environment. But the development of architecture is buffered against both genetic and environmental insults so that it reliably develops across the ancestrally normal range of human environments. Evolutionary psychologists do not assume that genes play a more important role in development than the environment does or that innate factors are more important than learning. Instead, evolutionary psychologists reject these dichotomies as ill-conceived.

Evolutionary Psychology Is Not Behavior Genetics Behavior geneticists are interested in the extent to which *differences* between people in a given environment can be accounted for by *differences* in their genes. Evolutionary psychologists are interested in individual differences primarily insofar as the differences are the manifestation of an underlying architecture shared by all human beings. Because the genetic basis of complex adaptations is universal and species-typical, their heritability (of the eye, for example) is usually very low, not high. Moreover, sexual recombination constrains the design of genetic systems so that the genetic basis of any complex adaptation, such as a cognitive mechanism, *must* be effectively universal and virtually species-typical (Tooby and Cosmides, 1990b).

Thus, the genetic basis for human cognitive architecture is universal and creates a universal human nature that can be precisely characterized. The genetic shuffle of **meiosis** (in animals, sex-cell production) and sexual recombination can cause individuals to differ slightly in quantitative properties that do not disrupt the functioning of complex adaptations. But two individuals do not differ in personality or morphology because one has the genetic basis for a complex adaptation that the other lacks. The same principle applies to human populations — from this perspective, separate and distinct human races do not exist.

In fact, evolutionary psychology and behavior genetics are animated by two very different questions — evolutionary psychology by, What is the universal evolved architecture that we all share by virtue of being humans? and behavorial genetics by, Given a large population of people in a *specific* environment, to what extent can *differences* between the people be accounted for by *differences* in their genes? The second question is usually answered by computing a heritability coefficient, for example, on the basis of studies of identical and fraternal twins. To the question, Which contributes more to nearsightedness, genes or environment? (an example of the second type of question), there is no fixed answer. The heritability of a trait can vary from one place to the next precisely because environments *do* affect development in interaction with genotype.

A *heritability coefficient* measures sources of *variance* in a *population*. For example, in a forest of oaks, to what extent are differences in height correlated with differences in sunlight, all else equal? The coefficient tells nothing about what caused the development of an *individual*. Suppose that 80% of the variance in height in a forest of oaks is caused by variation in the oaks' genes. This does not mean that the height of the oak tree in one's yard is 80% genetic. (Did genes contribute more to the oak's height than sunlight did? What percentage of the oak's height was caused by nitrogen in the soil? By rainfall? By the partial pressure of CO_2? What could any of these assertions mean?) When applied to an individual, such percentages are meaningless because all the factors are necessary for a tree to grow — remove any one, and the height is zero.

Joint Product of Genes and Environment Confusing individuals and populations has led many people to define the nature-nurture question in the following way: What is more important in determining an individual organism's phenotype, its genes or its environment?

A developmental biologist knows that this is a meaningless question. *Every aspect of an organism's phenotype is the joint product of its genes and its environment.* To ask which is more important is like asking, Which is more important in determining the area of a rectangle, the length or the width? or, Which is more important in causing a car to run, the engine or the gasoline? Genes *specify* how the environment can regulate the development of phenotypes; the specific details of the actual environment an individual organism encounters causes one or another of the regulatory outcomes to emerge.

Indeed, the developmental mechanisms of many organisms were *designed* by natural selection to produce different phenotypes in different environments. The fact that the environment controls the outcome is the product of natural selection having selected genes that create a particular developmental contingency between environment and outcome. For example, certain fish can change sex. Blue-headed wrasse live in social groups consisting of one male and many females. If the male dies, the largest female turns into a male. The wrasse are *designed* to change sex in response to a social cue — the presence or absence of a male.

With a causal map of a species' developmental mechanisms, one can change the phenotype that develops by changing its environment. Imagine planting one seed from an arrowleaf plant in water and a genetically identical seed on dry land. The one in water would develop wide leaves, and the one on land would develop narrow leaves. Responding to this dimension of environmental variation is part of the species' evolved design. But this result does not mean that any aspect of the environment can affect the leaf width of an arrowleaf plant. Reading poetry to it does not affect its leaf width. By the same token, it does not mean that it is easy to get the leaves to grow into any shape — short of using scissors, one cannot get the leaves to grow into the shape of the *Starship Enterprise*.

Unfortunately, people often have mystical ideas about genes and what they do. For example, people think that genes are the equivalent of what philosophers call *essences* — the hidden, unchangeable nature of something that inevitably gives rise to surface properties regardless of the environment in which the genes develop. But **genes** are simply regulatory elements, molecules that arrange their surrounding environment into an organism through an extraordinary long chain of causation, that can be disrupted or altered at any number of points. There is nothing magical about the process: DNA is transcribed into RNA; within cells, at the ribosomes, the RNA is translated into proteins — the enzymes that regulate development. There is no aspect of the phenotype that cannot be influenced by *some* environmental manipulation. It just depends on how ingenious or invasive one wants to be. If a human **zygote** (a fertilized human egg) is dropped into liquid nitrogen, it will not develop into an infant. If particles are shot at the zygote's ribosomes in just the right way, it may influence the way in which the RNA is translated into proteins. By continuing this process, one could, in principle, cause a human zygote to develop into a cactus or an armadillo. There is no magic inevitability here, only long chains of causality organized by natural selection to build functional structures within environments that resemble ancestral environments.

Present at Birth? Sometimes people think that to show that an aspect of the phenotype is part of our evolved architecture, one must show that it is present from birth, but this confuses an organism's initial state with its evolved architecture. Infants do not have teeth at birth — they develop teeth long after birth. Does this mean infants learn to have teeth or are teeth the result of socialization? What about breasts? Beards? One expects organisms to have mechanisms that are adapted to their particular life stage, such as the sea-squirt example discussed earlier — after all, the adaptive problems an infant faces are different from those that an adolescent faces. The human brain and mind matures along specific trajectories in all normal environments, and cognitive programs (such as the mechanism that causes romantic love) do not have to be present at birth to be part of our universal evolved architecture.

This misconception frequently leads to misguided arguments. For example, people think that if they can show that there is information in the culture that mirrors how people behave, *that* information is the cause of their behavior. Thus, if people see that men on television have trouble crying, they assume that their example is *causing* boys to be afraid to cry. This idea usually accompanies anthropological urban legends about hypothetical distant cultures in which practices were completely different. But which is cause and which is effect? Does the fact that men do not cry much on television teach boys to not cry, or does it merely reflect the way boys normally develop? In the absence of research on the particular topic, there is no way of knowing. (To understand this viewpoint, consider how easy it would be, by the normal logic of such arguments, to deduce that girls learn to have breasts: (1) breasts are absent at birth, (2) there is considerable peer pressure during adolescence to have breasts, (3) girls are bombarded with images of glamorous adult women with emphasis on breasts, (4) the whole culture reinforces the idea that women should have breasts, therefore, (5) adolescent girls learn to grow breasts.) In fact, an aspect of our evolved architecture can, in principle, mature at any point in the life cycle, and this applies to the cognitive programs of our brain just as much as it does to other aspects of our phenotype.

Is Domain-Specificity Politically Incorrect? Sometimes people favor the notion that everything is learned, by which they mean "learned via general-purpose circuits," because they think this idea supports democratic and egalitarian ideals. They think this view means that anyone can be anything. But the notion that anyone can be anything gets equal support, whether our circuits are specialized or general. When one talks about a species' evolved architecture, one is talking about something that is *universal* and *species-typical* — something all humans have, which is why the issue of specialization has nothing to do with democratic, egalitarian ideals — all humans have the same basic biological endowment, whether in the form of general-purpose mechanisms or special-purpose ones. If all humans have a special-purpose, language acquisition device, for example (see Chapter 6), all humans are on an equal footing when learning language, just as they would be if they learned language via general-purpose circuits.

Innate Is Not the Opposite of Learned For evolutionary psychologists, the issue is never learning versus innateness or learning versus instinct. The brain must have a certain kind of structure for learning to occur — after all, 3-pound bowls of oatmeal do not learn, but 3-pound brains do. To learn, there must be a mechanism that causes learning. Since learning cannot occur in the *absence* of a mechanism that causes it, the mechanism that causes learning must *itself* be unlearned, that is, innate. Certain learning mechanisms must therefore be aspects of evolved architecture that reliably develop across the kinds of environmental variations that humans normally encountered during their evolutionary history. Humans must, in a sense, have innate learning mechanisms or learning instincts. The interesting questions are, What are the unlearned programs? What are their design features, What is the structure of their procedures? Are they specialized for learning a particular kind of thing or are they designed to solve more general problems? These questions bring us back to Principle 4.

Specialized or General-Purpose? One of the few genuine nature-nurture issues concerns the extent to which a mechanism is specialized for producing a given outcome. Most nature-nurture dichotomies disappear when one understands more about developmental biology, but this one does not. For evolutionary psychologists, the important questions are, What is the *nature* of human universal, species-typical, evolved cognitive programs? and, What kind of circuits do humans *actually* have?

The debate about language acquisition brings the issue into sharp focus: Do the same hypothetical, general-purpose cognitive programs that cause children to learn (for example, to ride a bicycle) also cause children to learn language, or is language learning caused by programs that are specialized for performing the task? This question cannot be answered a priori. It is an empirical question, and the data collected so far suggest the latter (see Chapter 6; Pinker, 1994; and the chimpanzee language debate in Chapter 3).

For any given observed behavior, there are three possibilities: (1) the behavior is the product of general-purpose programs (if such exist), (2) the behavior is the product of cognitive programs that are specialized for producing that behavior, or (3) the behavior is a by-product of specialized cognitive programs that evolved to solve a different problem. (Writing, which is a recent cultural invention, is an example of the latter.)

More Nature Allows More Nurture There is not a zero-sum relationship between nature and nurture. For evolutionary psychologists, learning is not an explanation — it is a phenomenon *that requires explanation*. Learning is caused by cognitive mechanisms and to understand how learning occurs one needs to know the computational structure of the mechanisms that cause learning. The richer the architecture of the mechanisms, the more an organism is capable of learning: Toddlers can learn English but large-brained elephants or the family dog cannot because the cognitive architecture of humans contains mechanisms that are not present in that of elephants or dogs. Humans cannot learn to echolocate as well as bats, despite far larger brains, because humans lack circuits that are part of bat neural architecture. Furthermore, learning is not a unitary phenomenon. The mechanisms that cause the acquisition of grammar, for example, are different from the mechanisms that cause the acquisition of snake phobias. The same explanation can be applied to reasoning.

What Evolutionary Psychology Is Not For the reasons discussed above, evolutionary psychologists expect that the human mind contains a large number of information-processing devices that are domain-specific and functionally specialized. The proposed domain specificity of many of the devices separates evolutionary psychology from the approaches to psychology that assume the mind is composed of a small number of domain-general, content-independent, general-purpose mechanisms — the Standard Social Science Model.

The domain-specificity view also separates evolutionary psychology from the approaches to human behavioral evolution in which it is assumed (usually implicitly) that so-called *fitness-maximization* is a mentally (though not consciously) represented goal, and that the mind is composed of domain-general mechanisms that can figure out what counts as fitness-maximizing behavior in any environment — even evolutionarily new environments (Cosmides and Tooby, 1987; Tooby and Cosmides, 1990a; Symons, 1987, 1992). Most evolutionary psychologists acknowledge the multipurpose flexibility of human thought and action but believe it is caused by a cognitive architecture that contains a large number of evolved expert systems that can be combined in powerful ways.

Reasoning Instincts — An Example

In some of our research, we have been exploring the hypothesis that human cognitive architecture contains circuits specialized for reasoning about adaptive problems posed by the social world of human ancestors. In categorizing social interactions, there are two basic consequences humans can have on each other: *helping* or *hurting*, that is, bestowing benefits or inflicting costs. Some social behavior is unconditional, for example, one nurses an infant without asking the infant for a favor in return. But many social acts are conditionally delivered, which creates a selection pressure for cognitive designs that can detect and understand social conditionals reliably, precisely, and economically (Cosmides, 1985, 1989; Cosmides and Tooby, 1989, 1992). Two major categories of social conditionals are social exchange and threat — conditional helping and conditional hurting — carried out by individuals or groups on individuals or groups.

We (Cosmides and Tooby, 1992) initially focused on social exchange for four particular reasons:

1. Many aspects of the evolutionary theory of social exchange (sometimes called *cooperation*, *reciprocal altruism*, or *reciprocation*) are relatively well developed and unambiguous. Consequently, certain features of the functional logic of social exchange could be confidently relied on in constructing hypotheses about the structure of the information-processing procedures that this activity requires.

2. Complex adaptations are constructed in response to evolutionarily long-enduring problems. Situations involving social exchange have constituted a long-enduring selection pressure on the hominid line. Evidence from primatology and paleoanthropology suggests that human ancestors have engaged in social exchange for at least several million years.

3. Social exchange appears to be an ancient, pervasive, and central part of human social life. The universality of a behavioral phenotype is not a *sufficient* condition for claiming that it was produced by a cognitive adaptation, but it is suggestive. As a behavioral phenotype, social exchange is as ubiquitous as the human heartbeat. The heartbeat is universal because the organ that generates it is the same everywhere. This is also a parsimonious explanation for the universality of social exchange: The cognitive phenotype of the organ that generates it is the same everywhere. Like the heart, its development does not seem to require environmental conditions (social or otherwise) that are idiosyncratic or culturally contingent.

4. Theories about reasoning and rationality have played a central role in both cognitive science and the social sciences. Research in this area can, as a result, serve as a powerful test of the central assumption of the Standard Social Science Model: that the evolved architecture of the mind consists solely or predominantly of a small number of content-independent, general-purpose mechanisms.

The evolutionary analysis of social exchange parallels the economist's concept of trade. Sometimes known as reciprocal altruism, social exchange is an "I'll scratch your back if you scratch mine" principle. Economists and evolutionary biologists had already explored constraints on the emergence or evolution of social exchange using game theory, modeling it as a repeated set of interactions known as the *Prisoners' Dilemma*. One important conclusion was that social exchange cannot evolve in a species or be stably sustained in a social group unless the cognitive machinery of the participants allows a potential cooperator to detect individuals who cheat, so that they can be excluded from future interactions in which they would exploit cooperators (Axelrod, 1984; Axelrod and Hamilton, 1981; Boyd, 1988; Trivers, 1971; Williams, 1966). In this context, a cheater is an individual who accepts a benefit without satisfying the requirements on which provision of the benefit was made contingent.

Such analyses provided a well-formulated basis for generating detailed hypotheses about reasoning procedures that, because of their domain-specialized structure, would be well designed for detecting social conditionals, interpreting their meaning, and successfully solving the inference problems they pose. In the case of social exchange, for example, the procedures led us to hypothesize that the evolved architecture of the human mind included inference procedures that are specialized for detecting cheaters.

To test this hypothesis, we used an experimental paradigm called the **Wason selection task** (Wason, 1966; Wason and Johnson-Laird, 1972). For about 20 years, psychologists had been using this paradigm (originally developed as a test of logical reasoning) to probe the structure of human reasoning mechanisms. In this task, the subject is asked to look for violations of a conditional rule of the form: *If P then Q*.

Consider the Wason selection task presented in Figure 4.3. From a logical point of view, the rule has been violated whenever someone goes to Boston without taking the subway. Hence, the logically correct answer is to turn over the *Boston* card (to see whether the person took the subway) and the *cab* card (to see whether the person taking the cab went to Boston). More generally, for a rule of the form *If P then Q*, one should turn over the cards that represent the values *P* and *not-Q* (to see why, consult Figure 4.2).

If the human mind develops reasoning procedures specialized for detecting logical violations of conditional rules, the answer would be intuitively obvious. But it is not. In general, fewer than 25% of subjects tested spontaneously make this response. Moreover, even formal training in logical reasoning does little to boost performance on descriptive rules of this kind (Cheng et al., 1986; Wason and Johnson-Laird, 1972). Indeed, a large literature exists showing that people are not very good at detecting logical violations of if-then rules in Wason selection tasks, *even when the rules deal with familiar content drawn from everyday life* (Manktelow and Evans, 1979; Wason, 1983).

The Wason selection task provided an ideal tool for testing hypotheses about reasoning specializations designed to operate on social conditionals, such as social exchanges, threats, permissions, obligations, and so on, because (1) it tests reasoning about conditional rules, (2) the task structure remains constant even if the content of the rule is changed, (3) content effects are easily elicited, and (4) a body of existing experimental results already existed, against which performance on new content domains could be compared.

For example, to show that people who ordinarily cannot detect violations of conditional rules can do so when the violation represents cheating on a social contract would constitute initial support for the view that people have cognitive adaptations specialized for detecting cheaters in situations of social exchange. To find that violations of conditional rules are spontaneously detected when they represent bluffing on a threat would, for similar reasons, support the view that people have reasoning procedures specialized for analyzing threats. Our general research plan has been to use subjects' inability to spontaneously detect violations of conditionals expressing a wide variety of contents as a comparative baseline against which to detect the presence of performance-boosting reasoning specializations. By seeing what content-manipulations switch on or off high performance, the boundaries of the domains within which reasoning specializations successfully operate can be mapped.

The results of these investigations were striking. People who ordinarily cannot detect violations of if-then rules can do so easily and accurately when the violation represents cheating in a situation of social exchange (Cosmides, 1985, 1989; Cosmides and Tooby, 1989; 1992). This is a situation in which one is entitled to a benefit only if one has fulfilled a requirement, for example, "If you are to eat those cookies, you must first

FIGURE 4.3

A Wason selection task. The rule expresses familiar terms and relations drawn from everyday life.

Part of your new job for the City of Cambridge is to study the demographics of transportation. You read a previously done report on the habits of Cambridge residents that says: **"If a person goes into Boston, then that person takes the subway."**

The cards below have information about four Cambridge residents. Each card represents one person. One side of a card tells where a person went, and the other side of the card tells how that person got there. Indicate only those card(s) you definitely need to turn over **to see if any of these people violate this rule.**

Boston	Arlington	subway	cab

fix your bed"; "If a man eats cassava root, he must have a tattoo on his chest"; or, more generally, "If you take benefit X, you must satisfy requirement Y." Cheating is accepting the benefit specified without satisfying the condition on which provision of the benefit was made contingent (for example, eating the cookies without having first fixed your bed).

When asked to look for violations of social contracts of this kind, the adaptively correct answer is immediately obvious to almost all subjects, who commonly experience a "pop out" effect. No formal training is needed. Whenever the content of a problem asks subjects to look for cheaters in a social exchange — even when the situation described is culturally unfamiliar and even bizarre — subjects experience the problem as simple to solve, and their performance jumps dramatically. In general, 65% to 80% of subjects solve the problem, the highest performance ever found for a task of this kind. They choose the "benefit accepted" card (for example, "ate cassava root") and the "cost not paid" card (for example, "no tattoo"), for any social conditional that can be interpreted as a social contract and in which looking for violations can be interpreted as looking for cheaters.

From a domain-general, formal view, investigating men eating cassava root and men without tattoos is logically equivalent to investigating people going to Boston and people taking cabs. But everywhere this reasoning ability has been tested (adults in the United States, United Kingdom, Germany, Italy, France, Hong Kong, schoolchildren in Ecuador, Shiwiar hunter-horticulturalists in the Ecuadorian Amazon), people do not treat social-exchange problems as equivalent to other kinds of reasoning problems. Their minds distinguish social-exchange contents and reason as if they were translating the situations into representational primitives such as "benefit," "cost," "obligation," "entitlement," "intentional," and "agent." Indeed, the relevant inference procedures are not activated unless the subject has represented the situation as one in which one is entitled to a benefit only if one has satisfied a requirement.

Moreover, the procedures activated by social-contract rules do not behave as if they were designed to detect *logical* violations *per se*; instead, they prompt choices that track what would be useful for detecting cheaters, regardless of whether this happens to correspond to the logically correct selections. For example, by switching the order of requirement and benefit within the if-then structure of the rule, one can elicit responses that are functionally correct from the point of view of cheater detection but are logically incorrect (Figure 4.4). Subjects choose the *benefit accepted* card and the *cost not paid* card — the adaptively correct response if one is looking for cheaters — *no matter what logical category the cards correspond to.*

To show that an aspect of the phenotype is an adaptation, one needs to demonstrate a fit between form and function: One needs **design evidence**. There are now a number of experiments comparing performance on Wason selection tasks in which the conditional rule either did or did not express a social contract. These experiments have provided evidence for a series of domain-specific effects predicted by our analysis of the adaptive problems that arise in social exchange. Social contracts activate content-*dependent* rules of inference that appear to be complexly specialized for processing information about this domain. Indeed, they include subroutines that are specialized for solving a particular problem within that domain: cheater detection.

The programs involved do not operate so as to detect potential altruists (individuals who pay costs but do not take benefits). They are not activated in social-contract situations in which errors would correspond to innocent mistakes rather than to intentional cheating. And they are not designed to solve problems drawn from domains other than social exchange — for example, they do not allow one to detect bluffs and double crosses in situations of threat, nor do they allow one to detect when a safety rule has been violated. The pattern of results elicited by social-exchange content is so distinctive that we believe reasoning in this domain is governed by computational units that are domain-specific and functionally distinct, what we have called **social contract algorithms** (Cosmides, 1985, 1989; Cosmides and Tooby, 1992).

There is, in other words, design evidence. The programs that cause reasoning in this domain have many coordinated features that are complexly specialized in precisely the

---

Consider the following rule:

Standard version: *If you take the benefit, then you pay the cost* (e.g., "If I give you $10, then you give me your watch.")
If P then Q

Switched version: *If you pay the cost, then you take the benefit* (e.g., "If you give me your watch, then I'll give you $10.")

	Benefit Accepted	Benefit Not Accepted	Cost Paid	Cost Not Paid
Standard:	*P*	*not-P*	*Q*	*not-Q*
Switched:	*Q*	*not-Q*	*P*	*not-P*

FIGURE 4.4

Generic structure of a social contract. The standard and switched versions express the same social contract. In solving such tasks, subjects choose the *benefit accepted* card and the *cost not paid* card, regardless of which logical category these correspond to. On a *switched* social contract, the choices correspond to the logical categories *Q* and *not-P*. This is the correct answer if one is looking for cheaters, but it is logically incorrect. The logically correct answer is to choose *P* and *not-Q*, no matter what they stand for. But choosing the *cost paid* card (*P*) and the *benefit not accepted* card (*not-Q*) on a switched social contract does not allow one to detect cheaters.

TABLE 4.1 **Computational Machinery That Governs Reasoning About Social Contracts**

Design Features

1. Machinery includes inference procedures specialized for detecting cheaters
2. Cheater detection procedures cannot detect violations that do correspond to cheating (for example, mistakes in which no one profits from the violation)
3. Machinery operates in situations that are unfamiliar and culturally alien
4. Definition of cheating varies lawfully as a function of one's perspective
5. Machinery is as good at computing the cost-benefit representation of a social contract from the perspective of one party as from the perspective of another
6. Machinery cannot detect cheaters unless the rule has been assigned the cost-benefit representation of a social contract
7. Machinery translates the surface content of situations involving the contingent provision of benefits into representational primitives (for example, benefit, cost, obligation, entitlement, intentional, and agent)
8. Machinery imports conceptual primitives even when they are absent from surface content
9. Machinery derives implications specified by the computational theory even when they are not valid inferences of the propositional calculus (for example, "If you take the benefit, you are obligated to pay the cost" implies "If you paid the cost, you are entitled to take the benefit")
10. Machinery does not include procedures specialized for detecting altruists (individuals who have paid costs but refused to accept the benefits to which they are therefore entitled)
11. Machinery cannot solve problems drawn from other domains (for example, does not allow one to detect bluffs and double crosses in situations of threat)
12. Machinery appears to be neurologically isolable from more general reasoning abilities (for example, it is unimpaired in schizophrenic patients who show other reasoning deficits)
13. Machinery appears to operate across a wide variety of cultures (including an indigenous population of hunter-horticulturalists in the Ecuadorian Amazon)

Alternative Hypotheses Eliminated

1. Familiarity can explain social contract effect
2. Social contract content merely activates rules of inference of propositional calculus
3. Social contract content merely promotes (for whatever reason) clear thinking
4. Permission schema theory can explain social contract effect
5. Any problem involving payoffs elicits the detection of violations
6. Content-independent deontic logic can explain social contract effect

Source: Adapted from Cosmides and Tooby, 1992.

ways one would expect if they had been designed by a computer engineer to make inferences about social exchange reliably and efficiently. They are configurations that are unlikely to have arisen solely by chance. Some design features are listed in Table 4.1, as are a number of by-product hypotheses that have been empirically eliminated (Cosmides and Tooby, 1989, 1992; Cosmides, 1985, 1989; Fiddick et al., 1995; Gigerenzer and Hug, 1992; Maljkovic, 1987; Platt and Griggs, 1993; Sugiyama et al., 1995).

The focus of evolutionary psychologists on adaptive problems that arose in our evolutionary past has led evolutionary psychologists to apply the concepts and methods of the cognitive sciences to many nontraditional topics —cognitive processes that govern cooperation, sexual attraction, jealousy, parental love, food aversions, timing of pregnancy sickness, aesthetic preferences that govern appreciation of the natural environment, coalitional aggression, incest avoidance, disgust, foraging, and so on (Barkow et al., 1992). By illuminating the evolved cognitive programs that give rise to our natural competences, the research is allowing, for the first time, the systematic discovery and characterization of human nature. The process of mapping the cognitive programs has increasingly indicated that the most distinctive feature of humans as a species — high intelligence — is not the result of mechanisms designed for general problem-solving. Instead, human high intelligence appears to be contrived from a tapestry of modules, each with a different kind of structured problem-solving expertise, which in interactive combination endow people with a zoologically unprecedented ability to analyze, understand, and solve problems.

Acknowledgments. We would like to thank Martin Daly, Irv DeVore, Steve Pinker, Roger Shepard, Don Symons, and Margo Wilson for many fruitful discussions of these issues; William Allman for suggesting the phrase, "Our modern skulls house Stone Age minds," which is a very apt summary of our position; and Bill Schopf, whose active engagement with the broad sweep of evolutionary issues provided the occasion for this paper. We are grateful to the James S. McDonnell Foundation and National Science Foundation Grant BNS9157-499 to John Tooby for financial support during the preparation of this chapter.

REFERENCES

Axelrod, R. 1984. *The Evolution of Cooperation* (New York: Basic Books), 241 pp.

Axelrod, R. and Hamilton, W.D. 1981. The evolution of cooperation. *Science 211*: 1390–1396.

Baillergeon, R. 1986. Representing the existence and the location of hidden objects: Object permanence in 6- and 8-month-old infants. *Cognition 23*: 21–41.

Barkow, J., Cosmides, L., and Tooby, J. 1992. *The Adapted Mind: Evolutionary Psychology and the Generation of Culture* (New York: Oxford Univ. Press), 666 pp.

Baron-Cohen, S. 1995. *Mindblindness: An Essay on Autism and Theory of Mind* (Cambridge: MIT Press), 171 pp.

Boyd, R. 1988. Is the repeated prisoner's dilemma a good model of reciprocal altruism? *Ethol. Sociobiol. 9*: 211–222.

Cheng, P., Holyoak, K., Nisbett, R., and Oliver, L. 1986. Pragmatic versus syntactic approaches to training deductive reasoning. *Cognit. Psychol. 18*: 293–328.

Cosmides, L. 1985. Deduction or Darwinian algorithms? An explanation of the "elusive" content effect on the Wason selection task (Ph.D. Thesis, Cambridge: Psychology Department, Harvard University), 293 pp.

Cosmides, L. 1989. The logic of social exchange: Has natural selection shaped how humans reason? Studies with the Wason selection task. *Cognition 31*: 187–276.

Cosmides, L. and Tooby, J. 1987. From evolution to behavior: Evolutionary psychology as the missing link. In: Dupre, J. (Ed.), *The Latest on the Best: Essays on Evolution and Optimality* (Cambridge: MIT Press), pp. 127–159.

Cosmides, L. and Tooby, J. 1989. Evolutionary psychology and the generation of culture, Part II. Case study: A computational theory of social exchange. *Ethol. Sociobiol. 10*: 51–97.

Cosmides, L. and Tooby, J. 1992. Cognitive adaptations for social exchange. In: Barkow, J., Cosmides, L., and Tooby, J. (Eds.), *The Adapted Mind* (New York: Oxford Univ. Press), pp. 163–228.

Dawkins, R. 1986. *The Blind Watchmaker* (New York: Norton), 332 pp.

Fiddick, L., Cosmides, L., and Tooby, J. 1995. Priming Darwinian algorithms: Converging lines of evidence for domain-specific inference modules. *Ann. Meet. Human Behav. Evol. Soc.*, Santa Barbara, CA, p. 117.

Fodor, J. 1983. *The Modularity of Mind: An Essay on Faculty Psychology* (Cambridge: MIT Press), 145 pp.

Garcia, J. 1990. Learning without memory. *J. Cognit. Neurosci. 2*: 287–305.

Gigerenzer, G. and Hug, K. 1992. Domain-specific reasoning: Social contracts, cheating and perspective change. *Cognition 43*: 127–171.

Hirschfeld, L. and Gelman, S. 1994. *Mapping the Mind: Domain Specificity in Cognition and Culture* (New York: Cambridge Univ. Press), 516 pp.

James, W. 1890. *Principles of Psychology* (New York: Henry Holt), 458 pp.

Johnson, M. and Morton, J. 1991. *Biology and Cognitive Development: The Case of Face Recognition* (Oxford: Blackwell), 180 pp.

Leslie, A. 1988. Some implications of pretense for the development of theories of mind. In: Astington, J.W., Harris, P.L., and Olson, D.R. (Eds.), *Developing Theories of Mind* (New York: Cambridge Univ. Press), pp. 19–46.

Leslie, A. 1994. ToMM, ToBY, and agency: Core architecture and domain specificity. In: Hirschfeld, L. and Gelman, S. (Eds.), *Mapping the Mind: Domain Specificity in Cognition and Culture* (New York: Cambridge Univ. Press), pp. 237–267.

Maljkovic, M.J. 1987. Reasoning in evolutionarily important domains and schizophrenia: Dissociation between content-dependent and content-independent reasoning (Undergraduate Honors Thesis, Cambridge: Psychology Department, Harvard University), 123 pp.

Manktelow, K. and Evans, J.St.B.T. 1979. Facilitation of reasoning by realism: Effect or noneffect? *Br. J. Psychol. 70*: 477–488.

Markman, E. 1989. *Categorization and Naming in Children* (Cambridge: MIT Press), 250 pp.

Mineka, S. and Cook, M. 1988. Social learning and the acquisition of snake fear in monkeys. In: Zentall, T.R. and Galef, B.G. (Eds.), *Social Learning: Psychological and Biological Perspectives* (Hillsdale, NJ: Erlbaum), pp. 51–73.

Ohman, A., Dimberg, U., and Ost, L.G. 1985. Biological constraints on the fear response. In: Reiss, S. and Bootsin, R. (Eds.), *Theoretical Issues in Behavior Therapy* (New York: Academic Press), pp. 123–175.

Pinker, S. 1994. *The Language Instinct: How the Mind Creates Language* (New York: HarperCollins), 494 pp.

Platt, R.D. and Griggs, R.A. 1993. Darwinian algorithms and the Wason selection task: A factorial analysis of social contract selection task problems. *Cognition 48*: 163–192.

Spelke, E.S. 1990. Principles of object perception. *Cognit. Sci.* 14: 29–56.

Sugiyama, L., Tooby, J., and Cosmides, L. 1995. Testing for universality: Reasoning adaptations among the Achuar of Amazonia. *Ann. Meet. Human Behav. Evol. Soc.*, Santa Barbara, CA, p. 136.

Symons, D. 1987. If we're all Darwinians, what's the fuss about? In: Crawford, C.B., Smith, M.F., and Krebs, D.L. (Eds.), *Sociobiology and Psychology* (Hillsdale, NJ: Erlbaum), pp. 121–146.

Symons, D. 1992. On the use and misuse of Darwinism in the study of human behavior. In: Barkow, J., Cosmides, L., and Tooby, J. (Eds.), *The Adapted Mind: Evolutionary Psychology and the Generation of Culture* (New York: Oxford Univ. Press), pp. 137–159.

Tooby, J. and Cosmides, L. 1990a. The past explains the present: Emotional adaptations and the structure of ancestral environments. *Ethol. Sociobiol. 11*: 375–424.

Tooby, J. and Cosmides, L. 1990b. On the universality of human nature and the uniqueness of the individual: The role of genetics and adaptation. *J. Pers. 58*: 17–67.

Tooby, J. and Cosmides, L. 1992. The psychological foundations of culture. In: Barkow, J., Cosmides, L. and Tooby, J. (Eds.), *The Adapted Mind: Evolutionary Psychology and the Generation of Culture* (New York: Oxford Univ. Press), pp. 19–136.

Trivers, R. 1971. The evolution of reciprocal altruism. *Q. Rev. Biol. 46*: 35–57.

Wason, P. 1983. Realism and rationality in the selection task. In: Evans, J.St.B.T. (Ed.), *Thinking and Reasoning: Psychological Approaches* (London: Routledge), pp. 44–75.

Wason, P. 1966. Reasoning. In: Foss, B.M. (Ed.), *New Horizons in Psychology* (Harmondsworth, UK: Penguin), 474 pp.

Wason, P. and Johnson-Laird, P. 1972. *The Psychology of Reasoning: Structure and Content* (Cambridge: Harvard Univ. Press), 264 pp.

Williams, G. 1966. *Adaptation and Natural Selection* (Princeton: Princeton Univ. Press), 307 pp.

FURTHER READING

Barkow, J., Cosmides, L., and Tooby, J. (Eds.). 1992. *The Adapted Mind: Evolutionary Psychology and the Generation of Culture* (New York: Oxford Univ. Press), 666 pp.

Dawkins, R. 1986. *The Blind Watchmaker* (New York: Norton), 332 pp.

Pinker, S. 1994. *The Language Instinct: How the Mind Creates Language* (New York: HarperCollins), 494 pp.

Symons, D. 1990. A critique of Darwinian anthropology. *Ethol. Sociobiol. 10*: 131–144.

Williams, G. 1966. *Adaptation and Natural Selection* (Princeton: Princeton Univ. Press), 307 pp.

CHAPTER 5

EVOLUTION AND INTELLIGENCE—BEYOND THE ARGUMENT FROM DESIGN

Terrence W. Deacon*

THE PERSISTENCE OF TOP-DOWN EXPLANATIONS IN BIOLOGY

When the theory of natural selection was first presented to the scholars of the last century, many found it too implausible to believe. The incredulity of many great thinkers at the time, from brilliant biologists to articulate theologians, was based on a well-reasoned, common-sense understanding of the world; that is, left to chance, things tend to get less organized, not more. For millennia before Darwin, the same reasoning led Aristotle to criticize the "natural philosophy" of his contemporary, Empedocles, who argued that all natural processes are the actions of blind chance and that organisms arise out of the preservation of useful accidents (Aristotle's *Physics*). Aristotle easily found innumerable examples of end-directed design in nature that he felt could in no way be explained from such a minimalist perspective. But Aristotle was wrong, and only after more than twenty centuries of musing about this conundrum did we come to realize the power of the opposed conception for explaining biological phenomena. When the logic behind Empedocles' insight was rediscovered and given a more substantive interpretation by Darwin and Wallace, it revolutionized biology by providing an answer to this counterintuitive problem. The answer has become widely appreciated, not just by biologists, but by those of the general public educated in basic biology.

Nevertheless, despite the wide acceptance of natural selection as the major source of evolved adaptations, it remains a very counterintuitive notion. In a famous early nineteenth-century monograph, *Natural Theology*, William Paley (1802) described how it would stretch all credulity to imagine that something as complex as a pocketwatch,

*Department of Biological Anthropology, Boston University, Boston, MA 02215 USA

with its precisely aligned and interdependent gears and movement, could have been organized by anything other than an intelligent Designer. Not just its complexity, but the apparent purposive (that is, functional) design of its parts marks it as a product of intelligence and forethought. Because organisms appear far more complex and intricate in design and function than a pocketwatch, Paley argued that it is even more absurd to imagine that they could have arisen without the guiding hand of an intelligent designer.

Darwin, however, offered a theoretical apparatus to explain how, despite great improbability, it might nonetheless be possible to explain the spontaneous production of complex adaptation in entirely nonpurposive terms. Though even a century later there are still persons who blindly doubt the reality of biological evolution, the efficacy of the selection process that Darwin proposed to account for evolution has become a standard laboratory tool in molecular biology, information processing, and even engineering, and is used to spontaneously generate adaptive solutions to complex problems. Even if we were to discover that it was not, as we suspect, the major formative principle at work in shaping the course of life on Earth, this would not change the fact that it is the only serious contender for a natural process capable of generating such elaborate and finely adapted structures. So unquestionable is this mechanism, that (in a modern inversion of Paley's claim) many biologists now argue that whenever we encounter adaptive complexity in nature, the "default hypothesis" must be that it is a product of natural selection.

Yet Paley's perspective has not been totally eradicated from biological explanations. A core feature of this view is in fact central to much biological thinking, despite the fact that no serious biologist can now doubt the role of Darwinian processes in the evolution of organisms. The thorough rooting out of this idea — which I believe is a major source of confusion about the nature of biological processes even for professional biologists — is the purpose of this chapter.

Paley's original insight was motivated by the analogy of a living organism to a precisely designed machine: a watch. The Darwinian response has been to deny the necessity of intervention by intelligent planning and construction to produce such complex organization. In the modern view, Paley's designer has been replaced by natural selection. But the implicit comparison of an organism to a machine has remained. The Darwinian view is that organisms are put together *as if* they were designed. The products of evolution are still conceived of as clockworks — each part fashioned with respect to its contribution to the function of the whole organism. The adaptations are, in a sense, imagined as being accidentally designed with respect to some purpose, such as foraging, predator evasion, or reproduction.

The engineering analogy comes naturally. Designers often find themselves looking to nature for new solutions to structural problems, and everything from molecular diffusion systems and artificial neural networks have borrowed design principles from living counterparts. It is difficult to avoid the habit of thinking of organism structures and functions in engineering terms, given the elegance of some of nature's solutions. The aerodynamics of bird wings, the streamlining of fish and dolphin bodies, the ball-and-socket structure of shoulder and hip joints, and the countercurrent exchange arrangement of blood and water flow in fish gills all have an engineering elegance that human inventions have struggled to emulate or have serendipitously converged on. So one might defend the engineering analogy, as at least a heuristic convenience, especially since engineering analyses can also yield insights about organismal functions. Unfortunately, such analyses can also be misleading, which becomes more of a problem as the complexity of the system increases, such as in brains and intelligent behaviors.

The view of natural selection as a designer and of organisms as its designs is an eliminative program. Design is replaced by the retention of useful accidents. Purposive function is replaced by mechanistic operations that are merely correlated with a result, not chosen to achieve the result. Intentional adjustment to the demands posed by the environment is replaced by the blind implementation of prespecified behavioral and developmental programs encoded in genes. Hence, Richard Dawkins (1976), in his **paradigm**-challenging book *The Selfish Gene*, asks the reader to imagine multicellular organism bodies as replication machines — great lumbering robots designed by genes to ferry them from one generation to the next.

And, though cautioning his readers to take his **anthropomorphic** (human-based) turn of phrase when speaking of "selfish genes" as merely a rhetorical convenience, Dawkins succeeds in conveying the counterintuitive perspective in part by cleverly inverting the typical view of organisms as purposive and of genes as mere strings of molecules. He beguiles his readers into abandoning the natural **teleological** perspective at one level by replacing it with a provisional, merely metaphoric, alternative at another level.

Thus, ultimately, this line of reasoning suggests that the one presumed exception to this post hoc logic, human rational intelligent purposive behavior, may not be so rational or purposive after all. Intelligence might be understood as the evolution of robotic programming of a particularly complex and high level of flexibility, but a predesigned program, nonetheless. The presumed conclusion appears to be that such behaviors and experiences are ultimately reducible to collections of preprogrammed mechanisms in which all but superficial recombinations of prespecified elements are prefigured, which offers a final eliminative step. Even the apparent intentionality and purposiveness of thought processes appears to reduce to a form of predetermined mechanism.

This conceptual short cut, embodied in the engineering model of organisms, is a source of misleading intuitions about how living systems are assembled and how they operate. It is a short cut that continues to lead astray otherwise lucid biological analyses because it allows one to ignore how the differences in the means for achieving functional adaptation have contributed to the forms they produce. Beginning from this perspective, one is implicitly invited to imagine that the plans for organism designs are passed from generation to generation in coded sequences along the strands of genetic information-containing DNA that make up each species' **genomes** (the total complement of genetic information) and that the plans are decoded and used to determine the realization of the designs during **ontogeny**, like construction workers following a blueprint. Organism design is imagined to be embodied in a set of genetic assembly instructions that have been written and debugged by natural selection (as opposed to an all-intelligent Designer).

Yet we know that evolution produces organisms according to a very different modus operandi than does the purposive design of useful tools and other devices. Not only do these two very different processes arrive at solutions to the same problems in very different ways, they inevitably also produce very different kinds of solutions. One of the main reasons is that the assembly of organisms is **autonomous** (self-determined) rather than being under external, extrinsic control, which imposes very different design constraints than those that occur when the assembly process and assembly instructions are independently conveyed and executed by a designer. Both methods for producing adaptation may yield superficial resemblances in response to common problems and yet hide radically different logics of design and function. The question is, What is this difference in design logic?

What I hope to demonstrate in this chapter is that the design logic we tend to use to describe living adaptations often exactly *inverts* the reality, and that its assumption can be as misleading as Paley's view was for understanding the nature of organismal evolution. I do *not* want to dispute the Darwinian interpretation of the origins of **phylogenetic** adaptations. In fact, I want to propose a more far-reaching and radical approach, one that is ultimately *noneliminative*.

THE ONTOGENETIC INFORMATION BOTTLENECK

A common misconception — often implicit in evolutionary arguments — is that all the adaptive variety that an organism exhibits in its body functions and behaviors can be explained as a consequence of information imparted through genetic inheritance (with the exception of a little fine-tuning added by learning).

Changes in gene frequencies produced by natural selection are indeed capable of generating adaptations of remarkable power and subtlety, but genetic change is not the only way that adaptive biological structures and processes can be produced. In an effort to respond to the incompleteness of the gene selection models of evolutionary processes, critics of "strong selectionist theories," such as Stephen J. Gould (Gould and Lewontin, 1979), have suggested that we consider that some adaptation occurs at higher levels in evolution, such as at the population, species, and higher taxonomic levels. Gould pointed out that chance historical contingencies can also play an important role. Most evolutionary theorists, however, suggest that the conditions under which species-level selection could contribute a significant adaptive result are quite restrictive and do not apply to most populations.

Other alternative explanations for the origins of order in organism design have considered the role that intrinsic (internal) systemic constraints play in determining systemic regularities. For example, Stuart Kauffman (1992) has shown how complex patterning can emerge spontaneously simply as a result of the finite relationship within networks of interacting elements such as genes. The constraints on the range of possible patterns introduce biases into evolution, which may have had a significant effect on the ways genes determine cell types and body structures. The extent to which the constraints can account for a large fraction of the patterns of organism design and evolution is still unknown. I argue that the constraints do not exhaust the possibilities and that evolutionary adaptation accounts that are restricted to an analysis of gene frequencies and structural-functional consequences are missing an essential middle level of explanation.

One major limitation on the power of population genetic selection processes to produce adaptive variety in body structures is that the genome is a fairly inefficient mechanism for transmitting intelligent adaptations — those that involve highly flexible, indirect, complicated or unobvious solutions to adaptive problems. Genetic transmission is a limited transmission medium because cellular molecular transmission of information is constrained by distances, rates, and specificity of the chemical processes that underlie it. There are, as a result, tradeoffs in both information density and in packet size. The tradeoffs translate into upper limits on genome and cell size and into constraints on metabolism, reproductive rates, and the size of the fraction of genes that carry structural information. There are reasons to suspect that the genomes of vertebrates are near the upper limits. These limitations need to be considered in the context of the information required for adaptation.

With each level of organization (molecular, cellular, structural, functional) in a complex multicellular body there is an *exponential increase* in the amount of information necessary to control its construction and operation. In addition, the range of possible adaptive responses also grows exponentially with the complexity of the organism, including the complexity of its genome. As a result, a solution-by-solution, mutation-by-mutation, trial-and-error search conducted at the lowest level of design information (genes) is simply not capable of adequately sampling a useful range of alternatives (Kauffman, 1992).

Finally, with increasing design complexity, it becomes almost certain that random changes in structure result in a *decline* of function, not enhancement, which produces increasing inflexibility. For example, fiddling with even a single connection in a television set or a digital computer inevitably degrades function and usually produces catastrophic results. Theoretical evolutionary genetics has assumed that such an incalculably minimal likelihood of useful errors in a complex system can ultimately be explained by appealing to the law of large numbers: vast periods of evolutionary time and huge numbers of individuals.

In accepting a narrow view of Darwinism-restricted-to-genes we may have ignored its role in numerous other levels of supportive processes that are essential for a full account of how gene selection can succeed in arriving at adaptation. Even if the astronomical statistics could bear the weight of the multidimensional adaptive search across the many levels of variables involved, such a means for achieving complex adaptation would be irrelevant in the face of a more information-efficient means.

Indeed, it seems obvious that natural selection favors any mechanism that reduces such a massive sampling problem and can generate the requisite information with less information transmission "overhead." Instead of relying on immense time and intense selection to achieve immensely improbable modifications of genes that just happen to fit each new adaptive problem, *I think that organisms have evolved ways to circumvent this need* wherever possible. Lineages that evolved ways to effectively off-load or delegate this information-generation and information-transmission demand to subsidiary systems inevitably supplanted systems for which time and numbers were the only recourse.

For generations, biologists imagined organismal development and evolution as additive processes — new adaptations, new structures, and new functions added to previous ones and produced more organisms and more complicated organisms from the bacterium to the bat, from the **zygote** (fertilized egg) to the full-grown zebra. But this view of development and evolution is an anachronism. With very few exceptions, *the history of life is characterized by experiments with diverse rearrangements of the same few basic body plans.*

On average, vertebrates have larger genomes than do species in most other animal phyla (though most of the genome is noncoding DNA that few single-celled organisms can afford to allow to accumulate), but among vertebrates there is no trend suggesting that species with larger, more complex bodies and brains require more genes to produce them. Most estimates place the number of genes coding for protein products in mammals such as mice and humans at a number less than 100,000. This is probably a reasonable estimate for vertebrates in general, since mammals have genome sizes in the middle range of vertebrates (Figure 5.1).

This estimate is particularly significant for brain evolution in vertebrates. First, it suggests that few if any additional genes are involved in the production of huge human brains as opposed to tiny dwarf salamander brains. Second, considering that human brains probably contain over 100 billion neurons, each with thousands of connections, it suggests that the wiring instructions that could possibly be specified genetically are probably insufficient by many orders of magnitude to wire brains connection-by-

FIGURE 5.1

The distribution of genome sizes for the major groups of vertebrate animals, shown by logarithmic scale. Light shading indicates relatively smaller numbers of animals with genome sizes. Notice the greater range of genome sizes in geologically older lineages, including extremely large and small genomes in fish and amphibians (arrow). Differences are almost certainly the result of increases in noncoding and redundant DNA. All large genome species exhibit extremely low metabolism, slow maturation, dwarf species, and unusually large cell sizes, an indication of the costs of pushing the upper limits of genomic size, even if the vast majority remains untranslated.

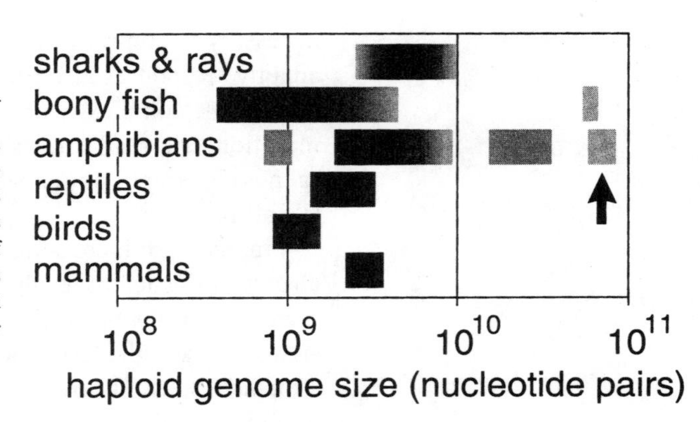

connection. And, of course, the extent of underdetermination of connectivity grows geometrically with size, which means that human brains, among the largest on the planet, must be among the least prespecified. All are reasons to suspect that the genetic instructions for building brains are not likely squandered on the micromanagement of individual neural connections and probably not even on the specification of tens of thousands of connections.

The foregoing brings up an even more enigmatic problem of brain design. If brain circuits are not well determined by genetic instructions, how is it possible that precise and predictable capabilities arise in evolution? How else could highly heritable behavioral capabilities evolve? In fact, brains are remarkably similar from individual to individual within a species and within different species within the same taxonomic orders, even when they differ greatly in size. Despite the growing information shortfall, large brains, such as human brains, are organized into many more distinguishable subdivisions than are small brains, and they are equally as well suited to operate their big bodies as are mouse brains for mouse bodies. *Where does the additional information come from, if not from genes?*

The answer requires a Darwinian but not a phylogenetic explanation. The process of brain development itself resembles a Darwinian process in important respects. Darwinian processes can spontaneously generate new useful information where none previously existed, and they need little prespecification to get under way. It is only natural, then, that the immense information gap in **ontogeny** (an organism's life history) should be filled in by the action of a kind of within-the-body natural selection that samples information that is implicitly present outside the genome. In this microcosm of evolution, the replication and growth of connections are the analogues of organism reproduction, competition for connection sites is the analogue of competition for resources and mates, and how well particular connections match the signal processing requirements of the body becomes the analogue of natural selection. Neural connection patterns that are appropriate to manage a complex body evolve to assume complementary niches of information processing, much the same as species evolve competitively and cooperatively to fill the complementary niches in an ecosystem (though the analogy is only partial, since ecosystems do not have their own consolidated means of replication). This evolution-like process for building brains allows brains to become adapted to the bodies they inhabit, as well as to their own internal constraints, with a minimum of preplanned design (Deacon, 1995).

DARWINIAN PHYSIOLOGY

Richard Dawkins takes the Darwinian alternative to its logical extreme by arguing that **Darwinism** is the *only* known mechanism for spontaneously producing adaptive complexity (Dawkins, 1986). There are two ways to interpret this claim. The first interpretation is what I have been describing as the standard evolutionary doctrine, namely, that all complex living processes are the results of biological evolution and are shaped in their designs by Darwinian selection. But there is a second, more radical interpretation possible, namely, that the complex adaptive processes that occur within and during the building of an organism are also *themselves* Darwinian-like processes, not merely their results. This second interpretation implicitly suggests that the details of complex adaptive responses are very minimally prefigured in genetic instructions, that they are created "online," so to speak, as the demand arises and at the corresponding level of organism structure. From this perspective, there is no need for a full plan for the organism to be encoded in the genome. What is required is something like the codes for a series of biasing mechanisms capable of coaxing otherwise spontaneous processes down predictable pathways.

Dawkins mostly intended the first interpretation, which is shown by his mechanistic account of organisms as mostly preprogrammed robots. But the two views are not at all at odds. The more encompassing perspective requires the traditional view as its limiting case. It might be caricatured by the phrase *"Darwinism all the way up"* — partially independent, nested Darwinian processes generate adaptive complexity at all levels of organism self-assembly and function.

The general utility of the selectionist model, beyond topics in the **phylogeny** of life, has been widely recognized. Extensive research has already shown that selection processes are the modus operandi for cellular-molecular mechanisms underlying immune response, **gamete** (sex cell) production, and aspects of neural development. However, even among biologists, few regularly consider how the basic elements of this explanation of spontaneously self-ordering processes apply to less remote aspects of organism function and what the broader significance of the logic of undesigned design might be. The reason for the ubiquitous role of evolution-like processes in widespread aspects of organism function is the generic nature of the conditions that can produce Darwinian processes, minimal conditions that can arise in a variety of contexts beyond population socioecology and genetics. These minimal conditions include:

1. Existence of self-replicating units of information whose replication is subject to low levels of spontaneous replication errors that arise independent of any function
2. Competition for resources that limit or enhance replication of these units
3. Consequent selective bias in replication rate and probability with respect to the differential functional consequences of the variations

The power of natural selection theory to explain the origins of adaptive complexity derives from its simplicity and inevitability given certain minimal conditions. It is no more than the statistics affecting replicative processes in the context of **entropy**. As the end of the twentieth century approaches, the scientific community has not only come to recognize the phylogenetic importance of the statistical mechanics of life but also is beginning to appreciate its ubiquity at every level of living function. Its widespread importance is a consequence of one simple fact: Darwinian selection is capable of spontaneously generating new adaptive structure and of achieving this by generating new information where none previously existed (unlike matter-energy, information can be created and destroyed). This is in fact what phylogenetic evolution demonstrates. Organisms take advantage of this fact by hierarchically embedding one Darwinian process within another to build their adaptations. It also provides a powerful way for organisms to amplify adaptive capacity and extend that capacity into more complex and unpredictable realms.

Hierarchic embedding of this type occurs spontaneously as higher-order, competitive-replicative processes are co-opted by lower-order ones. Within an organism, competition between gametes, selective replication of immune cells, and competition between growing **axons** (nerve-cell processes) for growth factors and **synapses** (nerve-cell connections) in the developing brain utilize variants of Darwinian processes to approach optimal adaptation and bridge the gap between gene-level processes and organism-level processes of selection. There may also be many more levels to this process than are now recognized. For example, many geneticists are beginning to suspect that the majority of the gene-to-gene interactions that control development and the molecule-to-molecule interactions that regulate cell metabolism also, at base, depend on the same sort of statistical mechanics of replication and competitive selection to produce complex adaptive effects.

The widespread employment of Darwinian processes in developmental and physiological processes derives from their ability to provide an autonomous source of adaptive responses. As both organic evolution and computer simulations of evolution demonstrate, they can arrive at nearly optimal solutions to adaptive problems without prespecification or monitoring of outcomes. The less that must be specified in detail the better, or rather, the more likely the processes are to be employed. Biological solutions produced by incorporating information that is spontaneously and reliably present inevitably supersede solutions that depend on information that must be passed on in an entirely prespecified form because systems that require detailed element-by-element design specifications are more susceptible to entropic influences, less flexible to perturbation and changing conditions, and more unlikely to adapt usefully in reasonable periods of time.

In contrast, organisms that have made the best use of spontaneous Darwinian mechanisms inevitably evolve faster and out-reproduce and out-compete mechanisms with more top-down design strategies. Moreover, hitting upon such a strategy is not unlikely. Darwinian processes arise spontaneously under a range of conditions that are common in living systems in which replicative processes are ubiquitous. Thus, the underlying evolutionary logic never needs to be specified or designed into an organism. It merely needs to be designed around, co-opted, and constrained by providing subtle biases to help shape the replication and selection processes involved.

COMPLICATIONS OF ORGANISM DESIGN

This view of organism design complicates how one thinks about comparative anatomy in a number of ways, for example, the notion of **homologous** organs (those of shared evolutionary derivation, such as human finger tips and horses' hooves or the cerebellum in mammal and frog brains). In classic theories, it was possible to distinguish between equating organs resulting from common descent of design information (**homology**) and organs resulting from common function because of convergence or parallelism (**analogy**). But this distinction becomes a problem when one considers ontogenetic Darwinian processes. The similarities between brains and brain systems in different individuals and different species must, therefore, be the result of a process analogous to convergent evolution in which highly similar selection forces act on homologous structures to produce similar organs.

The more directly homologous the substrates and the more highly constrained the selection conditions, the more similar the end product. Consider, for example, vertebrate pectoral fins and wings. Sharks, lobe-finned fishes, plesiosaurs, and dolphins converged on essentially the same fin design from somewhat different starting points (ray-finned fishes arrived at a similar solution but used different construction material). Similarly, pterosaurs, bats, flying lemurs, phalangers, and flying squirrels converged on essentially the same wing design (birds arrived at a similar solution, again using different construction material).

The design information, so to speak, was implicit in the preconditions of selection. The more similar the preconditions are, the more similar the result. The corresponding similarity of selection preconditions in development is, however, a distributed relationship among a great many body systems, which contributes a great deal of informational redundancy. The result is that, for the most part, incremental changes in one element have a subtle (seldom catastrophic) effect on the whole. Indeed, one recent support for this hypothesis comes from studies with so-called **knock-out breeds**, in which geneticists inactivate a specific gene and see what happens when the organism tries to develop. Often, even with apparently very important genes, knock-out organisms develop to resemble normal individuals because other systems have found ways to compensate, in part, for what was missing (important advances are most often made when one discovers a major reorganizational consequence). There appears to be a sort of hierarchy to this plasticity. Events that occur early in development (as a result of genes that are turned on early) and whose effects tend to ramify in all later stages are more highly conserved (which implies that they are under stronger genetic selection) than events that occur later.

BRAINS ADAPT TO BODIES

There is an evolutionary tendency for building organisms as evolutionary processes *within* evolutionary processes as shown by brain evolution and brain development. This most complex product of evolution is the ultimate expression of amplification of adaptation. The expression of this logic is most explicit with brains. It is also where intuitions about design, purpose, and mechanism most easily become confused, where mechanistic and purposive models come into direct conflict, and where common-sense intuitions find both accounts wanting. These qualities make brain development an ideal place to start to trace such a thoroughly Darwinian logic of organism design. Ultimately, the problem of explaining the apparent intelligence behind evolutionary adaptations and of explaining the evolution of intelligence as an adaptation can be seen as the same problem and may have similar explanations. As more serious attempts are made to discover the links between brain evolution, brain development, and cognitive processes, it becomes increasingly important to appreciate the related processes at intermediate levels between genetic evolution and cognition.

If brains were designed the way one designs watches or computers, flexibility from generation to generation and evolutionary adaptation in the long run would be nearly impossible. The brains of large mammals, such as humans, are some of the most complex objects that have ever existed. With so many millions of interconnected, interdependent parts, significant modifications of one component must be correlated with complementary modifications of innumerable other interdependent components to avoid catastrophic disruption of system-wide functions. This is one very good reason why no one recommends redesigning computers or television sets by randomly modifying connections or parts in millions of devices and testing to see which mutations work best. Aside from the obvious waste involved, the chances that *any* of the millions of possible changes produce enhanced function are essentially zero. Technological progress is the result of redesign in which the engineers involved must pay very close attention to all the detailed ways that their innovations interact with one another and the other parts of the device. For example, redesigning and upgrading complex systems such as microprocessors and jet airliners requires tens of thousands of engineers contributing thousands of hours of checking and rechecking both how their parts work and how they interact with parts designed by others. Even then, surprises tend to emerge when the whole system is first assembled. These designs are not nearly as complex as simple organisms and their brains. Living things are not like designed devices because the engineering approach is far too cumbersome to have evolved.

A close look at brain development shows the levels of Darwinian processes involved. To predict the outcome of a Darwinian process, one needs to know two things: (1) the sources of bias affecting the growth and replication of the basic units of selection (specifically, neural connections) and (2) the invariant features of the context in which the growth occurs that provide the competitive milieu and selection biases that determine which units will develop and persist and which will not. The parallels are not difficult to recognize in brain development.

In the process of cell production, one can see clear evidence of selection processes at many levels of the nervous system (Purves and Lichtman, 1980, Cowan et al., 1984). Some of the first examples discovered involved the production of neurons directly controlling muscles: **motor neurons** (Purves, 1988). It was noticed that during development these output neurons were produced in greater abundance than they are in mature individuals. Sympathetic **ganglia**, whose neurons projected to the smooth muscles of the viscera, and spinal motor neurons that projected outputs onto limb muscles seemed to go through a culling process as the organism matured. In experiments on early-stage embryos of frogs and chicks, it was discovered that the extent of the culling could actually be increased or decreased by modifying the peripheral organs to which they projected axons. By removing their targets, more cells were induced to die off; by grafting additional organs (for example, a supernumerary limb), fewer cells were eliminated (Figure 5.2A). Apparently, the cells were initially overproduced and found themselves in competition for resources that were somehow provided by the peripheral target structures. These resources turn out to be both molecular (like analogues to food) and structural (like analogues to protective burrows).

Early in development, the output branches of motor neurons (axons) grow somewhat exuberantly and nonspecifically and end up overlapping one another on the same muscle cells. Competition ensues to make contacts (synapses) with the cells. In the end, only one axon wins and only those with stable synapses seem to provide the molecular signals necessary to keep the source neuron alive. Branches of axons that fail to establish connections die back, so to speak, and cells that fail to establish any stable connection die entirely (see Figure 5.2B).

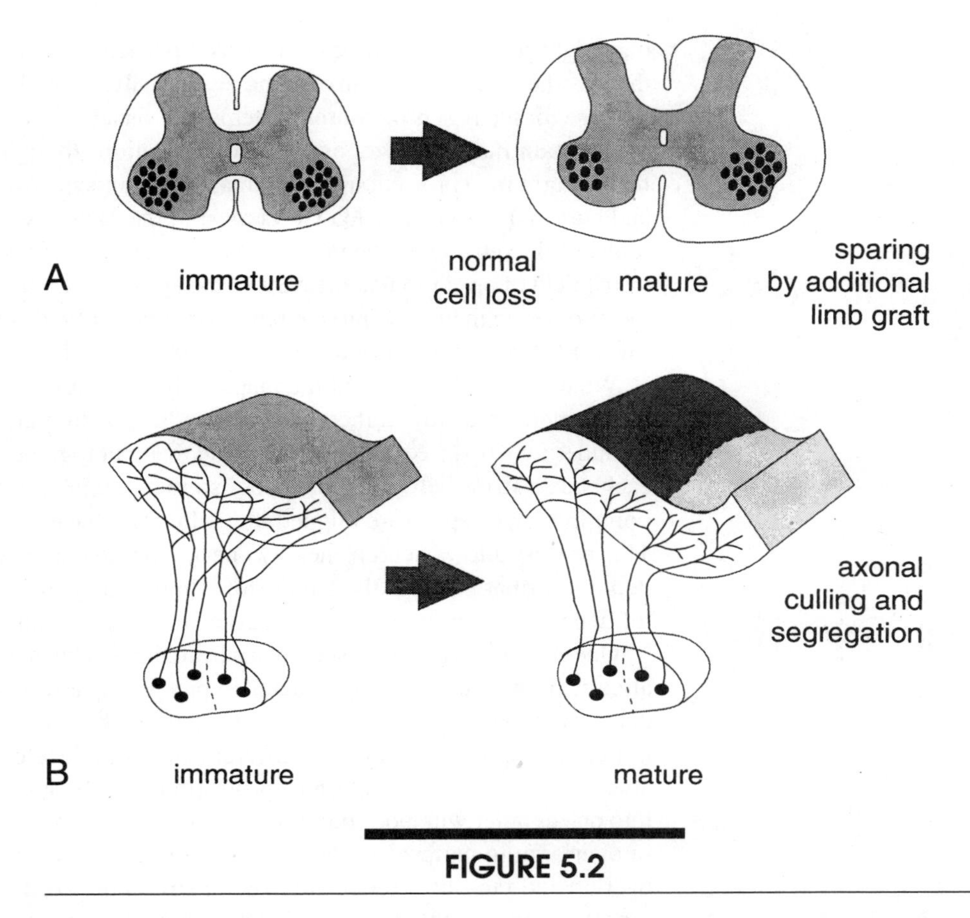

FIGURE 5.2

Two basic Darwinian-like processes involved in central nervous system development. Neuronal and axonal overproduction and nonspecificity and subsequent selective cell death and axonal culling are the major means of matching neural populations and functional topology of different linked systems to one another during development. The selectional basis of this process can be demonstrated in cases where additional targets are added or normal targets are reduced. (**A**) Grafting an extra limb on a young embryo can reduce the extent of motor neuron cell death in the ventral horn of the corresponding side of the spinal cord. (**B**) Experiments comparing topography of connections between immature thalamus (ovoid object, *below*) and cortex (sheet-like body, *above*) to their mature counterparts have shown that early projections are more numerous and diffuse than are adult projections and have demonstrated that axonal competition for target territories results in highly topographic innervation patterns.

Selection-mediated cell death, thus, turns out to provide a precise mechanism for matching the sizes and distribution of populations of neurons to the sizes and distribution of muscle masses in the rest of the body. From an evolutionary point of view, there need be no correlated change in neural cell production or cell distribution to match the changes in muscle size and distribution that have resulted from selection for different modes of locomotion (Deacon, 1990; Katz and Lasek, 1983; Purves, 1988; Wilczynski, 1984).

The same logic turns out to be utilized throughout the developing brain, not just for motor systems but also for sensory systems and even intrinsic systems. For example, in different vertebrates the direction the two eyes face may differ from almost 180 degrees (as in many fish and hooved mammals) to almost 0 degrees (as in owls and humans). In species in which the visual fields overlap, there is the possibility of using the nearly redundant information to aid in depth perception, but given the range in possible overlap, one might suspect that it needs to be accomplished differently in different brains. Indeed, the way the connections map onto the visual analyzers in the brain is quite complex. As is the case for many **sensorimotor** systems, the visual projections into the central analyzers of the brain maintain a topographic organization (though somewhat distorted, as are many world map projections). In animals with binocular overlap (for example, monkeys and humans) each half-retina views most of the same visual field as the same side half-retina of the other eye. The projections into the brain of an adult split according to their visual field of origin so that the parts of the eye that view the left visual field cross over in the midline on their way into the brain. Both visual fields map onto the **cerebral cortex** (the folded gray matter sheet that covers each half of the right side of the forebrain).

What is remarkable about the maps is that they are both separated and overlapping in a complicated way in the visual cortex. It is as though each map had been cut into meandering strips (like zebra stripes) and then put together in a single map, interposing each side's strips between the other's, so that points on a strip that represent the same point in viewed space are aligned immediately adjacent to one another (Figure 5.3A). This process allows neurons nearby one another to compare signals and thereby extract depth information from the slight shifts in disparity that result from the way distance influences the convergence or divergence of lines of sight.

One might imagine that such complicated map organization, requiring such precise alignment, would require very detailed prespecification. Evidence to the contrary was discovered more than 2 decades ago when it first became possible to trace the course of individual input connections at different stages in development. What was found was an early, rather messy, pattern of projections in which the two eyes' maps "diffused" into one another with poor point-to-point precision. Axons tended to branch and fan out in overlapping patterns in the visual cortex. But during development, the degree of overlap and fan-out is reduced via competition between axons, and many (of what in hindsight one might label as misdirected and nonspecific) branches are selectively eliminated to produce the final precisely sculpted pattern. Much more has been learned since then about the level of initial biases that help axons approximate their initial trajectories and about the nature and extent of competitive processes involved, but the basic logic underlying all is Darwinian logic.

The power of selection processes to produce such a precise, complex pattern has been most forcefully demonstrated by a parallel pattern produced in the frog visual system, but in this case by a very abnormal experimental manipulation. Law and Constantine-Paton (1981) grafted an additional (third) eye onto frog embryos' heads to study the effect on axon growth to the brain. In frogs, the major visual analyzer is not the cortex (because frogs do not have a brain structure that is exactly comparable to cerebral cortex) but rather a paired midbrain structure called the **optic tectum** (like the cortex, a sheet-like structure), which contains the visual maps. Frogs' eyes, like those of other amphibians and of reptiles and fish, do not view any significant degree of overlapping visual space, and each retina projects its map in total to opposite sides of the optic tectum.

However, "triclops" frogs (three-eyed, experimentally produced frogs) experience much visual overlap. The third eye, placed above and adjacent to one of the existing eyes, views a large part of the same region of space as its near neighbor. When the experimenters examined the projection pattern in the optic tectum, they discovered that these two eyes projected interdigitated stripe maps, analogous to mammal binocular maps, in this very different brain structure (see Figure 5.3B). What is so remarkable about this finding is that although evolution has never given rise to frogs with binocular vision of this sort — and thus there never before had been a need for a mechanism to precisely interdigitate visual

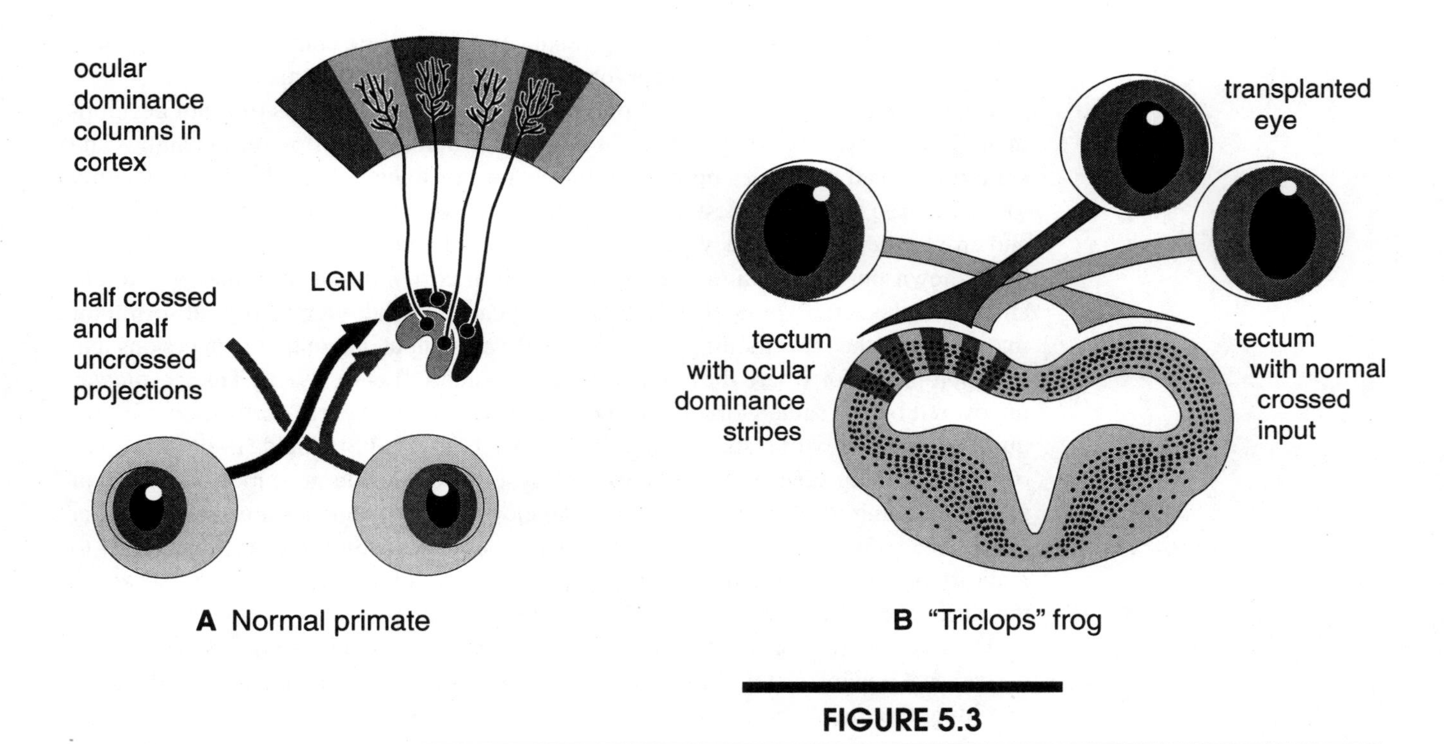

FIGURE 5.3

Role of axonal competition in generating precisely topographic connection patterns was initially suggested by observing the precision of retinotopic projections from the eyes to the visual processing centers in the brains of mammals and frogs. (**A**) One of the most complex topographic patterns is associated with the interdigitation of retinotopic maps from the two eyes in the cerebral cortex of binocular mammals such as cats and primates (shown here). Those portions of the retinas of each eye that view the same position in the visual field project to the same region of visual cortex, but to adjacent strips or ocular dominance columns. (**B**) An analogous pattern was discovered to be produced in the frog visual tectum (a midbrain structure not a cortical structure) in response to transplantation of a third eye, which introduced axonal competition biased by the correlated input patterns from eyes that viewed the same point in space. This binocular pattern almost certainly has no evolutionary precedent in frogs.

maps — it emerged spontaneously, without evolutionary precedent, precisely because of a similarity in the processes of selection. This is an expression of convergent developmental selection, analogous to the convergent evolution of fins and wings but one that produces much more precise convergence from seemingly very different origins.

These examples beg the question of exactly how the competitive processes work. The answer is only incompletely known and involves many mechanisms. Most agree on two critical components: (1) growth factors provided by the recipient (postsynaptic) cells are necessary to maintain patent connections and axons compete for these factors; and (2) what determines whether an axon will be able to hold onto a synaptic connection with respect to competing axons vying for the same target cell has something to do with correlating activity patterns that favor the axons that tend to fire in synchrony with one another and also the target cell staying connected to the same target cell (Purves, 1988).

The end result is that subtle spatial and timing biases in initial connectivity contribute to the relative synchrony or dyssynchrony of signals converging on any particular target, and the biases become amplified by the progressive action of millions of signals and selective elimination events that ensue. Thus, some initially rather subtle biases that can be controlled developmentally by some rather generic growth processes are capable of adjusting the development of both large-scale and microscale connectional patterning. In the end, they produce intricate appropriately detailed circuits to match.

It now appears that this general principle is at work at all levels of brain development. In fact, the remarkable pattern of maps for different sensorimotor modalities and submodalities that divides up the cerebral cortex of mammals probably is also only loosely biased by genetic design and yet comes to exhibit a remarkable interindividual and cross-species consistency. For example, studies by O'Leary (1992) and colleagues have shown that transplanting immature cerebral cortex tissue from one position to another in the cerebral mantle of rats does not cause the later-appearing sensorimotor maps to track the repositioning. Rather, they develop input and output connections that are appropriate for where they end up (Stanfield and O'Leary, 1985). The reason was uncovered by the same workers in studies that showed that early cortical outputs are quite nonspecific with respect to their targets (O'Leary and Stanfield, 1989). All areas of cortex initially send outputs to nearly all types of cortical targets, mostly by way of extra branching of their axons. Later in development the branches are pruned so that visual areas project only to subcortical visual processors, auditory areas project only to auditory processors, and motor areas project only to motor structures. Thus, where in the cortex the projections originate does not really matter.

The same is true for inputs to cortex that determine which information each sector processes. Growing chunks of fetal cortex and **thalamus** (an ovoid mass of brain nuclei providing the major source of inputs to cortex) in tissue culture has shown that which combinations of cortical and thalamic regions are grown together does not matter; each interconnects as well as any other (Molnçr and Blakemore, 1991; Yamamoto et al., 1992). Moreover, studies in which the growth of developing inputs to the thalamus from various sensory systems have been interrupted early in development demonstrate that the remaining inputs can recruit the thalamic regions (nuclei) that were supposed to be the targets for the missing inputs (Frost and Metin, 1985; Sur et al., 1988).

The shift in thalamic input patterns is also passed on to the cortical representation of these nuclei. The cortical tactile auditory region can, for example, become visually responsive because of visual inputs taking over the auditory nucleus of the thalamus (the **medial geniculate nucleus**). A curious natural experiment demonstrates this. The blind mole rat (*Spalax*) has only vestigial eyes. In its brain, the visual nucleus (the **lateral geniculate nucleus**) gets most of its input from auditory subcortical sources, and there is a corresponding shifting of functional boundaries in the rest of the thalamus and the recipient areas of cortex to match (Doron and Wollberg, 1994). Where other rodents have visual cortex, the blind mole rat has auditory and somatic sensory cortex, not because the one was eliminated and the others added in any modular sense, but because in the competition for space, the displacements of connection patterns at lower levels produced ramifying effects at all levels. Thus, cortical areas' inputs and outputs are both competitively determined along with the patterns of connections within cortex. The pattern generation process is entirely systemic and distributed.

BRAIN TRANSPLANTS

One consequence of the fact that most information for wiring brains is produced, not inherited, in each generation, is that most aspects of species' brain differences are likely not specified in any detail. The same genetic and molecular information can serve very different purposes in different parts of the brain and in very different brains. Some of the most convincing evidence for this cross-species generality comes from experiments in which **chimeric brains** were produced by transplantation techniques. Immature neural cells harvested before the point in development at which neurons have extended fragile axons and dendrites can be transplanted from one brain to another, both across ages (from fetal to adult) and across species. By placing immature neurons from one stage or species into another one can experimentally probe for the relative influence of intrinsic and extrinsic signals for development.

As part of an effort to develop alternative fetal cell sources for transplantation treatments for neurodegenerative diseases, my colleagues and I at McLean Hospital (Belmont, MA) studied the growth of fetal cells from pig brains in the brains of adult rats and monkeys (we have recently had the opportunity to follow this process in human clinical trials). The Darwinian features of neural developmental processes had given us reason to predict that the signals that controlled neural maturation and connectivity in the different species would likely not be very dissimilar and thus might allow cross-species transplants to function nearly as well as same-species transplants (so long as immune rejection could be held at bay, as was accomplished with the same sorts of drugs used for other types of organ transplants). It also gave us a chance to observe exactly how the different species' cells interact.

One interesting discovery was that the developmental clock that decides when cells move on to succeeding stages of differentiation and growth seems to be intrinsically controlled. Pig cells matured at the same rate whether in pig brains or rat brains. This finding turned out to be a boon, both for the experiments and for understanding features relevant to their clinical application. Adult brains are highly resistant to axonal growth, one reason that brain damage is seldom repaired spontaneously. We still do not know why the inhibitory mechanisms are activated, but they slow axon growth by at least one order of magnitude, which means that fetal rat neurons implanted into an adult rat brain often cannot grow their axons to reach appropriate targets. They reach the end of their growing period without establishing synapses; lacking appropriate support, the axons, and sometimes the cells, die. But pig neurons stay immature and grow more than five times as long as do rat neurons; as a result, even at their slow rate of growth through the stubbornly resistant adult brain tissue, pig axons grow to reach targets that rat axons cannot (Deacon et al., 1994; Isacson et al., 1995).

Remarkably, pig neurons taken from one region of the fetal pig brain and implanted into an adult rat brain grew their axons to the rat brain target structures that were homologous to the normal pig brain targets for those neurons. Cortex cells grew to striatum and midbrain, striatal cells grew to midbrain but not cortex, and midbrain cells grew to striatum and cortex, and so on (Isacson and Deacon, in press). Growth into these targets did not appear to be topographical, but generic, and occurred even if the transplants were placed in odd sites within the host brain. Most important, the newly established connections replaced functions in rats that had lost these functions because of loss of some of their own neurons. Thus, we have been able to approach human clinical therapy using pig cells with some confidence that transplantation will succeed. Indeed, early results are quite promising (Figure 5.4).

The implications of transplantation for evolution are that even incredibly species-specific neural functions, such as echolocation or language processing, have probably been achieved, not by specific circuit changes encoded by specific genes for those functions but by using the old developmental information slightly modified by systemic effects and generic biasing of global brain relationships. Pig cells may be playing

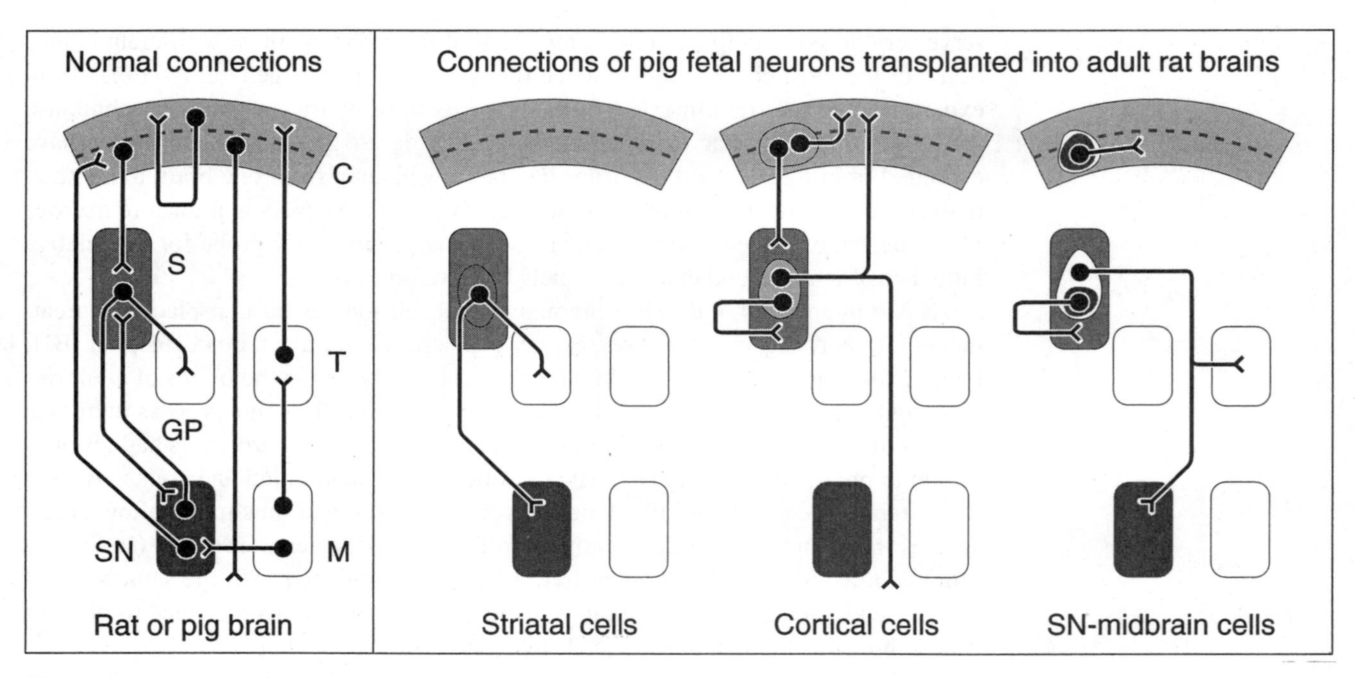

FIGURE 5.4

Conservatism of initial axonal guidance mechanisms demonstrated by cross-species transplantation studies using fetal neurons whose axons can be followed as they grow into the host brain. (**A**) Normal axonal connections of structures used in transplantation experiments. C, cerebral cortex; S, striatum; GP, globus pallidus; T, thalamus; SN, substantia nigra; MB, other ventral midbrain structures adjacent to the SN. Dashed line in the cortex indicates Layer 4 distinction between infra- and supragranular layers (cortical projection neurons below this division are the source of most striatal and brain-stem projections, whereas upper layers project mostly to other cortical areas).

(**B**) Results of three transplantation experiments in which the cells from different fetal pig brain regions (striatum, neocortex, and ventral midbrain, respectively) have been transplanted into the striatum or cerebral cortex of adult rats and allowed to mature. Fetal pig striatal neurons (homotopic site) grow axons selectively into the globus pallidus and the substantia nigra (*left*). Fetal pig neocortical neurons (ectopic site) grow axons selectively into the striatum, cerebral cortex, and the corticospinal tract leading out of the forebrain (*middle*). Fetal pig midbrain neurons (including SN precursors and other midbrain precursors) (ectopic sites) grow different classes of axons into different targets. Those that become dopaminergic neurons (as in the SN) selectively grow into the striatum if transplanted into the striatum and into the deep layers of cortex if transplanted into the cortex; non-dopaminergic neurons from other parts of the ventral midbrain grow axons to very different targets, avoiding striatal gray matter and growing into the thalamus and back to the midbrain (*right*). To reach these normal targets, pig axons must make appropriate use of rat guidance cues.

a role in supporting improved language functions in human patients, even as you are reading these words! For this reason, understanding how such a uniquely human function arose in the first place may not require any highly specific genetic explanation. There may be no intrinsically prespecified language circuits in the human brain, only circuits that have inherited this function as a consequence of a doubly Darwinian process.

Does this mean that there is no modularity and no prespecificity at any level in brains? Is the whole affair a Darwinian competitive free-for-all? Though I believe that the determination of organism design is Darwinian even to the molecular level, this belief does not preclude the evolution of highly localized modular systems, nor does it exclude the possibility of significant species and individual differences that have a precise genetic basis. However, understanding that these differences also arise from Darwinian mechanisms during development may help us to gain a clearer understanding of the conditions that must be met for such specificity to result.

Size plays a critical role. Many small worms and related creatures (for example, leeches) have nervous systems in which the neuron-by-neuron identity and wiring are highly predictable, so much that it is possible to individually name (or number) specific neurons in specific ganglia and brain regions. In small body sizes, individual neurons are much more like separate organs subserving distinct functions, which precludes the kind of statistical specification that goes on in species in which each neural organ is composed of hundreds to billions of neurons, all processing similar information. But even in very small species with countable neurons, programmed cell death and systemic selective processes seem to be at work; they are just more highly and easily constrained by their context.

The level at which higher-level brain functions are specified in different species has recently been studied directly by putting whole chunks of developing brains into other species to replace corresponding chunks removed from the brains. One of the most interesting examples of this experimental approach used developing chicks and quails. It turns out to be possible to remove a section of the neural tube in a very young Japanese quail embryo and replace the section with the correspondingly removed section of a chick neural tube, as long as this procedure is done at a very early stage when neurons have not yet been produced and would thus be damaged by the process.

In one surprising experiment, Evan Balaban and colleagues (1988) swapped quail for chick midbrain regions. Remarkably, the chicks with the quail midbrains continued to develop and hatched able to behave quite appropriately except that they exhibited certain characteristic stereotypic movement patterns and calls of quails. (Unfortunately, both for them and for science, immune rejection quickly ensued, so the chimeric birds never had a chance to mature.) These brain inserts had matured, established extensive and appropriate interconnections with surrounding host brain structures, made appropriate functional links (even though the quail brain structures were much smaller than the corresponding chick brain structures would have been), and yet retained certain whole functional programs characteristic of the donor species.

Clearly, both highly integrated and complex cross-species interconnectivity can develop as well as some level of species-specificity of function. They are not incompatible. In comparison with the pig xenograft procedures described above, however, one important difference is the implantation of precursors of whole, intact brain regions. Thus, positional and growth cues within the midbrain region itself were appropriate to quail brains. The scaffolding that served as a source for selection and growth biases affecting neural connections *within* the midbrain were laid down by quail ontogeny. It is likely that the connections necessary for producing the specific stereotypic bird behaviors are almost entirely contained within this segment of the brain (as has also been suggested by experiments involving electrical stimulation of and damage to structures located within the midbrain). A key factor in the evolution of such preprogrammed functions is almost certainly their embodiment within circuits that are localized within a relatively confined part of the brain, so that they are relatively insulated from systemic effects during development. They are modular in both structure and function. Highly constrained localized structures and modular function probably go hand-in-hand in this regard.

A characteristic feature of such **modular** functions is also that they are almost entirely automatic and fixed in their production. We would not normally call them intelligent behaviors, though they clearly may be components in higher-order intelligent behavior patterns. It is important to recognize, however, that modularity of function also emerges from the development of distributed systems, like that involved in skill learning. When humans learn a motor skill, for example, playing a scale with one hand on the piano, what begins as a process that involves a great deal of attention to sensory feedback and active inhibition of other movement tendencies eventually becomes automated so that it can be performed as though activating an external mechanical device. Analysis of the distribution of brain activity before, during, and after learning simple mental skills clearly shows that less of the brain is required to be involved as a skill becomes more automatic. Both the genetic evolution of such innately automated skills and the production of their facultative counterparts during life probably follow parallel paths of reduction of representation in the brain, but a great deal is yet unknown in both cases to trace either process in detail.

HOW FLY GENES BUILD MAMMAL BRAINS

The partial segmental independence of the development of connectivity within whole segments of the brain, such as that exemplified by the quail-chick transplants, is a reflection of a more general design strategy in embryology, which represents a middle level in the hierarchy of Darwinian processes that intercede between genes and connectional specification of the brain. At the level at which global features of brain organization are determined, some crucial biases shaping later cellular and connectional selection processes are introduced; therefore, it is to this level that many characteristic species differences trace their ontogenetic origins.

The discovery of the molecular genetic (DNA-RNA) basis for transmitting and decoding genetic information revolutionized biology more than any other discovery since Darwin. Recently, the discovery of a class of genes that regulate the large-scale patterns of regional differentiation of early embryos has revolutionized the study of development (McGinnis and Kuziora, 1994). Probably the most unexpected feature of these genes is that they not only do the major work of sculpting bodies and brains, but they are shared almost unchanged in species as divergent as flies and humans (Holland et al., 1992).

The genes share the common attribute of being able to regulate the activation of other genes because they include one or more molecular base sequences that code for protein structures that bind directly to DNA. Many of these genes are called **homeotic genes** because they partition the embryo into segmental divisions that exhibit similar (homological) structures.

Examples of segmentally homologous structures are found in the vertebral column and similar forelimb and hindlimb structures. The brain and spinal cord initially develop from a neural tube that runs the length of the embryo. Shortly after the tube forms (by an infolding of the embryo surface), it is partitioned by the localized expression of specific sets of homeotic genes into segmental regions that eventually become the major divisions of the brain and spinal cord. What is remarkable about the genes is that almost identical genes are activated in the fly body during its development (and in other animals as disparate as worms and snails) in the same order and relative positions as in vertebrate embryos, presumably including humans (Figure 5.5).

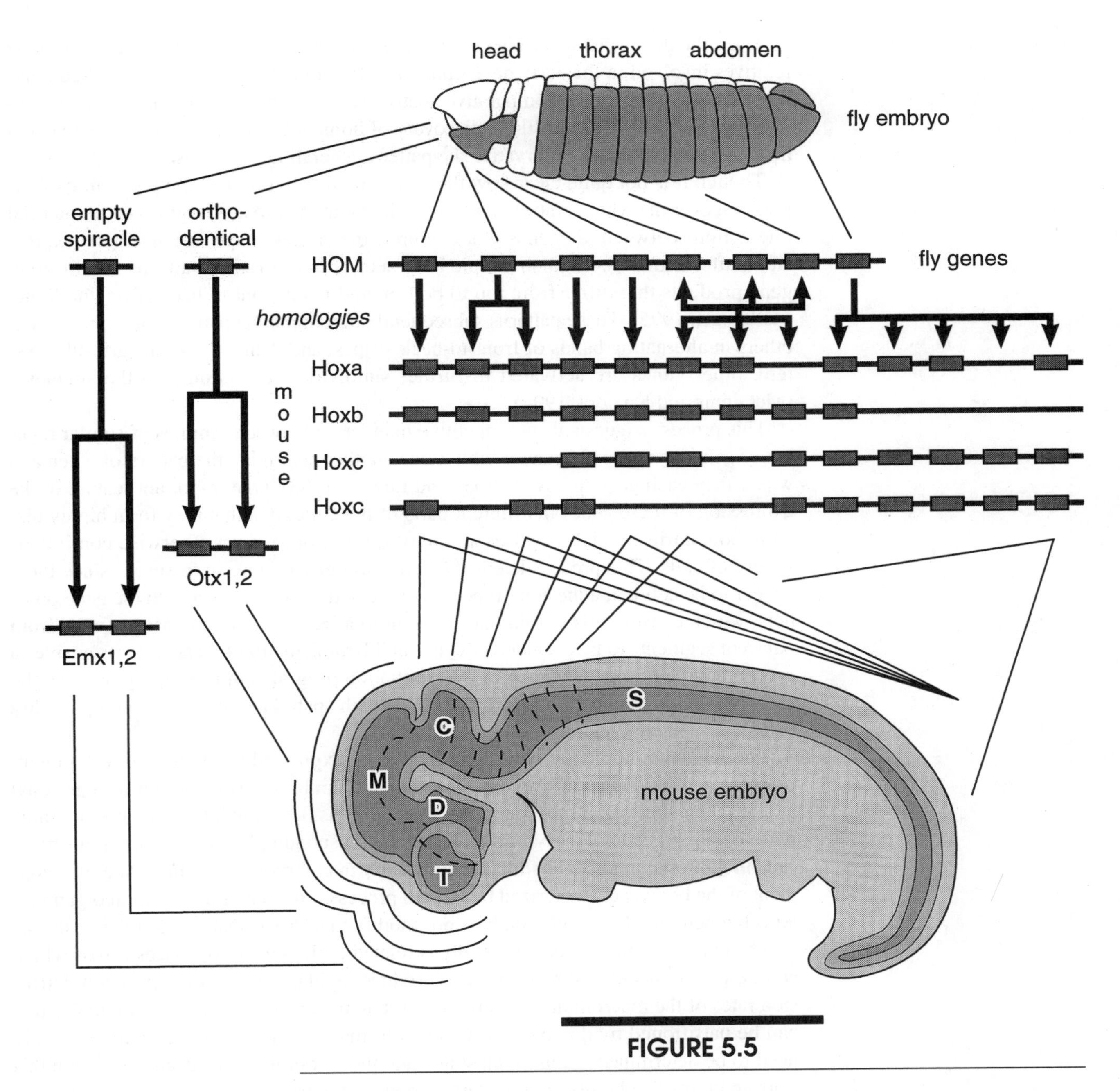

FIGURE 5.5

Mapping of segmental patterns of corresponding homeotic gene expression in the ectoderm of fruit fly larvae (*above*) and the mouse embryonic nervous system (*below*). Genes are depicted as rectangles on a line (a continuous length of chromosomal DNA); fly genes are indicated in the top row, corresponding mouse genes in the lower rows. Arrows indicate putative homologies between the genes, including two duplication homologies: duplication of the whole string of *HOM* genes to produce four clusters of *Hox* genes and duplication of individual genes (split arrows). Lines from genes to body regions roughly indicate point of expression. Note that in contrast to the exclusive segmental expression in the fly embryo, the mouse genes are expressed in partially overlapping regions (indicated by regions within the diverging lines for *Hox* genes and by contour lines for *Otx* and *Emx* genes). Both produce segmental patterns, but in the vertebrate pattern, because of the overlap, the segments are defined by having different combinations of genes expressed. The temporal order of expression of these genes also corresponds to their relative positions on the chromosome so that those *HOM* genes expressed in the head end are first and those at the tail end of the abdomen are last. *Otx* and *Emx* genes and their fly homologues are expressed in a mirror-image order.

Thus, the genes that determine early stages of human brain development have close relatives involved in the early development of fly brains! The similarity verified a hypothesis offered by many comparative anatomists: Most bilateral animals share a common logic of body plan. Until the discovery of homeotic genes, however, no one could have guessed just how conservative the pattern-generating mechanisms are.

Though it is not quite clear how the genes interact to produce the patterning of expression over the whole embryo body, it is clear that the process requires combinatorial interactions between the genes. For example, the earliest stages involve gene expression patterns that distinguish simple body axes, concentration patterns for different gene products that differ from top to bottom and from front to back (Boncinelli and Mallamaci, 1995). These patterns subsequently determine the patterns of expression of others in alternating bands of front-to-back stripes, and then, within this grid-like system, other genes are activated to further subdivide and distinguish the segments (McGinnis and Kuziora, 1994).

This process suggests that competitive interaction between families of similar regulatory genes vying to activate or inhibit one another determine the pattern of when and where they each get expressed (Hoey and Levine, 1988). One important feature of the expression of these genes in different parts of the embryo is that they form highly discrete boundaries, which imposes a kind of digital logic on an otherwise continuous sphere of cells. This process keeps the cell's progeny in line, so to speak, since there appear to be cell-cell adhesion properties expressed in conjunction with the gene products, which causes cells within one segment to adhere and congregate but cells from adjacent segments to be excluded (Keynes and Krumlauf, 1994). The process enables a certain degree of independence of differentiation programs in nearby regions, yet the serially homologous organization guarantees both spatial and molecular compatibility between segmental regions.

Though some homeotic genes continue to be expressed throughout development, most are activated at comparatively early stages in embryo formation (**embryogenesis**) and are then shut off; almost certainly their roles as determinants of large-scale morphology are played out early in the process. Correspondingly, the formation of organs (**organogenesis**) tends to be initiated quite early in embryogenesis; most later development of the body is characterized by growth processes that enlarge the initial organ systems but neither add to their numbers nor modify their basic spatial interrelationships.

The reason for this process probably is related to the limited distances across which competitive molecular interactions can produce regular patterns of expression. Diffusion rates of the macromolecules probably set an upper limit to the size of embryo that can be partitioned by this means. As a result, major divisions of the brain and body need to be determined at this small stage, and the rest of growth extrapolates from this starting point. The highly similar starting point for extrapolative growth probably contributes much of the regularity in allometric (size-shape) structural relationships among related species, even though not all structural divisions grow to the same extent.

Segmentation of the neural tube and growth of corresponding brain regions follows this logic. The expression domains of early regulatory genes in the head end of the neural tube identify regions that go on to become distinct sorts of neural tissues. These domains are initially laid down in a sort of front-to-back sequence of zones called **neuromeres**. In the brain stem and spinal cord, these simple segments are identified by the sequential expression of a set of genes called ***Hox* genes** (named for the homeobox gene cluster in fruit flies in which the genes were first identified), which are arranged in four **paralogous** (homologues in the same genome) linear clusters (Kappen and Ruddle, 1993; McGinnis and Kuziora, 1994). The order of the *Hox* genes on the chromosome and the order of their temporal and spatial expression from the base of the midbrain to the tail end of the neural tube are the same in flies and mammals (and probably all lineages in between). Most vertebrates differ by having two to four partially redundant clusters, the significance of which is not yet known (see Figure 5.5).

The region in front of the midbrain-brainstem transition is somewhat less linear. A number of other regulatory genes are expressed in a quasisegmental pattern and obey many of the same divisions of **midbrain**, **diencephalon**, and **telencephalon** (brain regions progressively forward along the neural tube), but they are not arranged in as regular a pattern as are the genes in more caudal (tail end) regions. Using a number of gene expression patterns and correlating them with morphological landmarks, Rubenstein et al. (1994) nevertheless identified a sequence of serially arranged **prosomeres** (neuromeres of the **prosencephalon**, which is the name for the undifferentiated forebrain at early stages), each of which is distinguished from adjacent ones by very distinct boundaries of gene expression. However, within the forebrain there seem to be many very distinctive gene-expression divisions between dorsal and ventral regions, which suggests that one might need to also think of these as separate half-prosomeres. At present, there is no single logic that explains this topology and there is considerable controversy about whether a simple continuation of segmental organization is valid for this brain region. However, most agree that these unambiguous divisions mark boundaries between regions that later develop into distinctive brain regions with distinctive cell types and functions.

Two small families of regulatory genes appear to follow a temporal and spatial expression pattern in the forebrain in a sort of mirror image of *Hox* gene expression along the brain stem and spinal cord. These families are identified as ***Emx*** and ***Otx*** (named, respectively, after fly counterparts — empty spiracle and orthodentical — discovered first in fly heads and then used to fashion molecular probes to look for mammal homologues). The two paralogues of each are expressed in a back-to-front, Chinese-box-type pattern (one within the domain of the other) in the order *Otx2* to *Otx1* to *Emx2* to *Emx1* (see Figure 5.5), with *Otx2* essentially defining most of the forebrain ahead of the brain stem and *Emx1* covering the smallest region, confined only to dorsal telencephalon, which becomes the neocortex (Simeone et al., 1992; Boncinelli et al., 1993).

Recent work with frogs suggests that the domain of *Otx2* expression can be enlarged or reduced either by increasing its representation in all cells (by genetically inducing extra expression in all cells from an earlier time point) or by inhibiting it with a differentiation factor called retinoic acid (Boncinelli and Mallamaci, 1995). When *Otx2* expression is augmented, it correlates with an *overdevelopment* of the head end of the neural tube; when its expression is inhibited, there is *reduction* of the head end. This finding, again, is consistent with a type of competitive expression process and also suggests a possible genetic basis for shifts in the segmental relationship between brain and body — the types of shifts that distinguish primates from most other mammals and distinguish humans from other primates (Deacon, 1991, 1995).

To date, there is scant evidence that the neocortex is subdivided by different homeotic gene expression domains, which is consistent both with its relatively uniform cellular structure and its multipotentiality during development. This finding should not be surprising from an evolutionary and developmental perspective because only in relatively recently evolved (particularly terrestrial) vertebrates has this region of the brain become especially prominent. At the time in development that most homeotic genes are expressed, the entire telencephalon is one of the smallest divisions, perhaps composed of only two prosomeric regions; thus, its few subdivisions are consistent in scale with those in other brain regions. The vast proportion of its disproportionate expansion occurs later in development in terrestrial vertebrates (especially in birds and mammals) after the highly conserved genetic partitioning events have taken place.

The distinctions and divisions between homeotic gene-expression domains appear to play a major role in directing early generic axon growth. Early axonal path-finding tends to show abrupt growth effects that differ across neuromere boundaries; genetic or transplantation manipulations of the position of the early gene-expression territories can produce corresponding redirection of axonal growth (Figdor and Stern, 1993; Kessel, 1993).

The regional differences in molecular activity in the brain probably provide the basis for crude, target-directed growth that gets axons to their destination in preparation for synaptic competition within the brain. Persistence of some of the homeotic guidance and specificity cues may be the source of generic species-general growth patterns that guide appropriate axon growth in xenografted (transplanted) brains.

Homeotic gene-expression patterns may even play some role in introducing more subtle biases. For example, O'Leary (1992) recently showed that the **retinotopic** map-patterning of projections from the eye to the tectum in chicks is influenced by the graded expression pattern of a gene called *engrailed* (also first found in flies). **Retino-tectal** axons grow in different directions in the tectum, depending on the concentration of engrailed protein expression and whether the axons originated from one sector of the retina or another, which may help explain why divisions of the cerebral cortex depend so much on Darwinian processes to be specified. This part of the brain is an expansion of a single homeotic segmental division and shares common axon guidance cues and cell types throughout its tangential extent.

The exception that proves this rule appears to be that *Otx* gene expression also distinguishes deep cortical layers from superficial ones (Frantz et al., 1994). This finding is correlated with the fact that although different sectors of cortex are not distinguished by different connection affinities, the different layers within each cortical region have target specificities distinct from one another and are shared in common with the same layer cells in other regions. The deepest layer is Layer 6 and its neurons only project to thalamic targets. Next, Layer 5 neurons project to the most distant subcortical targets, including the striatum, midbrain, brain stem and spinal cord, but not the thalamus. All superficial layer neurons project to cortex either locally or more distantly; only very few, deep Layer 3 neurons find their way to the nearest noncortical target, the striatum. This exception provides further evidence for a conserved homeotic determination of initial connectivity.

In summary, all the genetically specified tissue differences and axon target specificities of vertebrate brains appear to be determined by a very conserved pattern of intercellular gene interactions at a very early point in embryogenesis. But although all vertebrate brains share most of the same major divisions and exhibit similar subdivisional architecture, there is nevertheless considerable variability among major classes. Only in mammals, for example, is there a thick, many-layered cerebral cortex. Thus, in mammals, the system is not totally conserved. Very likely, researchers will uncover homeotic gene-expression differences that correspond to the brain structure differences between vertebrate classes, but at present, too few cross-species comparisons exist to allow even an educated guess as to what the differences might be — except that they probably involve the same highly conserved genes (or additional duplicates of them) that perhaps are shifted in expression domains or timing.

WHAT MAKES HUMAN BRAINS HUMAN?

On the basis of what is known about human brains compared with other species' brains, one can venture reasonably educated guesses about the changes in ontogenetic processes that caused humans to deviate from more typical ape patterns. First, there is no evidence of addition of new parts, no extra neuromeric segments, no new kinds of brain structures, no special-purpose devices plugged in, and almost certainly, no new kinds of homeotic genes. Nevertheless, human brains have more of *something* making the brains large. The human brain is a primate brain in every respect — except that it is peculiarly large for the body. But human brains are not simply primate brains scaled up, the way one might enlarge a photograph.

Comparisons of brains from large and small mammals in general, and large and small primates in particular, demonstrate remarkably predictable cross-species size relationships among the parts of the brain. For example, plotting the sizes of two different brain regions with respect to one another in a spectrum of species of primates (Figure 5.6) shows a very orderly **allometry** (metric of differential growth) linking most regions together. Knowing the size of one major brain region, one can predict the size of most others with remarkable accuracy, on the basis of the shared trend in the group, across a wide range of sizes. Not coincidentally, the regions that produce the most highly correlated trends correspond roughly to homeotic segmental divisions, which suggests that a common mitotic growth process may be applied to all and that the trends reflect the way this scalar factor extrapolates the initial partitioning of brain regions.

But it is not a simple extrapolation because different parts scale according to slightly different trends. For example, a larger fraction of the brain in larger-brained primates is forebrain (especially cerebral cortex) and cerebellum; less is midbrain, brain stem, and spinal cord (Finlay and Darlington, 1995). This finding suggests that with increasing growth away from a common, highly similar embryonic stage, slight differences in extrapolated growth have an increasingly greater impact on relative proportions of brain structures in different species.

What determines the size of a given brain structure is, of course, the number of cell divisions that take place because homeotic processes have determined the ultimate cell fates for that region. This process is largely a function of time spent in **mitotic growth** (via body cell division), since in mammals, at least, there seems to be a relatively stable rate at which brains grow in the womb (Sacher and Staffeldt, 1975; Deacon, 1995; Deacon, 1996). One crude hint of how this might influence proportional differences comes from comparing the trend differences with the timing of some of the underlying developmental changes. Structures that exhibit a net **positive allometry** (that is, that tend to be proportionately larger in larger species) also tend to be the latest structures to mature, namely, the cerebral cortex and cerebellar cortex. As noted earlier, there is also a brain stem-to-telencephalon sequence of homeotic gene expression; thus, the latest forebrain regions to mature are also the latest to be carved up by the genes. Perhaps this consistent difference in reaching comparable maturational stages is responsible for a few additional cell divisions that, in larger species, get progressively extrapolated to greater proportional differences in the size of the corresponding brain structures. Whatever the reason, the allometric regularities again indicate how remarkably conserved and predictable is the neural developmental process.

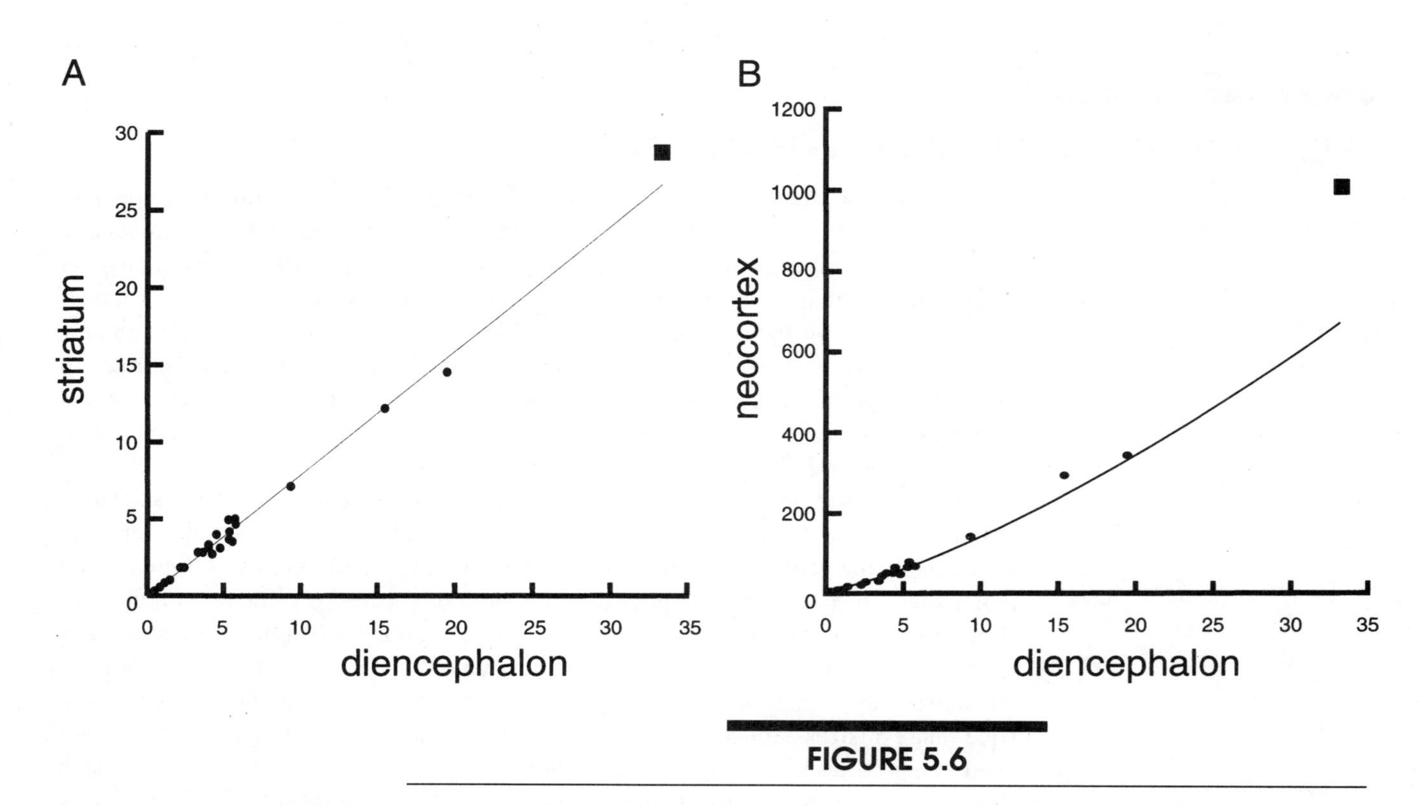

FIGURE 5.6

Two examples of proportions of major segmental divisions of the human brain as compared with divisions in other anthropoid primates. Other primates' brain regions generally obey remarkably correlated growth patterns with respect to one another (indicated by their tight fit on the regression line, computed without data for humans, denoted by solid square). (**A**) Anthropoid trend in the relative size of the basal ganglia with respect to the size of the diencephalon (including thalamus and hypothalamus) can be accurately extrapolated into the human size range (solid square). (**B**) The anthropoid trend does not predict the size relationship between the neocortex and the diencephalon, despite the fact that the structures are extensively interconnected. Because these shifts in proportions influence (bias) axonal competition during development, it is likely that human brains exhibit significant shifts in adult axonal connection patterns as compared to other primates.

However, human brains do not fit the predictions in certain interesting ways. In each plot of comparative brain structure allometries, the human point is plotted for comparison with the other primate trends. Sometimes (as shown in Figure 5.6A) the human values are well within the range extrapolated from other primates, but in other cases (Figure 5.6B), the human values deviate considerably from the prediction. If the predictability of the patterns reflects conserved neuroontogenetic processes, the human deviations indicate a break from the norm.

Human brains are not just scaled-up primate brains; they differ at the level of whole neuromeres, which implies that there may be some homeotic change that distinguishes human brain development from all other primate brain development. The *Otx* and *Emx* genes offer relevant candidates for such change since the structures that are disproportionately larger in human brains are initially defined by these genes' expression domains. What the change may be is not predictable at this time, except that it probably involves these or related gene expression events.

One can, however, predict some consequences from what is already known about the Darwinian nature of brain development. The shift in proportions in major brain regions should also affect the axonal competition for connections throughout the brain, just as would the addition of a large number of additional sensory receptors or extra limbs. This is the neurological analogue of introducing a major new resource into an ecosystem that preferentially benefits one species. The result is that all selection relationships are altered to some degree. Unlike the blind mole rat or the frog with an extra limb, this change is centrally located and thus has its major influence of connectional development from the inside out. I have argued elsewhere (Deacon, 1990; 1995; 1996) that the change would tend to influence sensory systems relatively little since they would be strongly constrained by their peripheral inputs — but that it would increase cortical output control over numerous subcortical systems, including motor systems involved in laryngeal and mouth movements. It would also alter the relationships between cortical areas because most cortical connections are with other cortical areas.

These are clues to exactly what differences in neural function were selectively favored in our ancestors. Was it simply increased intelligence in some generic sense? The appearance of significant rearrangement of neural proportions and the ramifications this has for connectional reorganization suggests something quite different. It suggests that there has been a change in another level of Darwinian processes: a cognitive level. It suggests that the balance of neural computations has been biased so that some systems within the brain have a significant competitive advantage in determining the course of the outcome of moment-by-moment neural activity patterns over others. From a Darwinian perspective, it represents a major innate shift in cognitive selection conditions that favors patterns of neural activity produced in the more expanded brain regions over those from the less expanded ones: a shift in the ecology of mind (to borrow a phrase from the famous late-nineteenth-century British naturalist Gregory Bateson). Certainly the one most distinctive cognitive faculty of humans, language, reflects such a shift, and not just a change in mental capacity or the implementation of a set of inborn grammar-decoding instructions. The puzzle is figuring out how the two are related.

EVOLUTION TURNED INSIDE OUT

In every generation, it seems, the most sophisticated technologies are appropriated to serve as metaphoric models for the mysterious working of brain processes. Within a century, models have shifted from hydraulic systems to telephone switchboards, and now, to digital computers (which have had the reverse honor of being attributed brain functions such as memory and intelligence). Even the computer hardware-firmware-software distinction has found its way into the metaphors of mental processes as an analogue to neural connections-innate abilities-learned abilities, though any similarities are extremely superficial. The algorithmic and computer metaphors, like all good analogies, have nonetheless sharpened the precision of models of mental processes.

One cannot only be explicit about predictions of the models, the predictions can be implemented in computer simulations of mental computing to see how they behave in a variety of contexts. The models are troublesome only if they are overinterpreted — if one mistakes the map for the territory. A few decades ago it was legitimate to withhold judgment about whether the processes of neural representation resembled **algorithmic processes**, that is, explicit symbolic codes and subroutines that were explicitly stored, retrieved, and executed to run sensory analyses or motor movements in much the same way as machine instructions determine a specific sequence of signal-processing transformations or revolutions of a stepper motor controlling a robotic arm.

It made sense to wonder whether innately predisposed response patterns were encoded and stored in the genome and then reloaded into the fetal neural memory banks, like uploading the operating system into RAM at machine startup. Thus, an innate **universal grammar**, specified as a set of rules for making certain classes of symbol transformations, could appear as a plausible model of human language acquisition predispositions.

As researchers now begin to understand the details of the genes to brain translation, they face a very different sort of logic. If one takes as metaphors of mind the biological processes that underlie mental processes and the logic that goes into the construction of brains, a very different set of predictions about the logic of mental processes emerges — one that is not well modeled by computer metaphors precisely because computers and their software traditionally are designed, not evolved.

Psychologists have long been comfortable (some would say too comfortable) with comparisons of high-level mental operations, such as learning, with evolutionary processes. For example, superficially, trial-and-error learning processes resemble evolutionary processes. Alternative responses are viewed as competing organisms; variations in response tendencies are viewed as mutations, and pleasurable or aversive outcomes are seen as natural selection that determines which responses are most likely to be repeated in the future.

The analogy of learning to natural selection has been recognized almost from the time Darwin's ideas became public. Indeed, even before Darwin, the analogy was implicit in the **Lamarckian** conception of evolution, and writers such as Herbert Spencer made it the core of his "Theory of Psychology." However, because of this similarity, the role of learning has also been a source of confusion in evolutionary theory, as Spencer and Lamarck well demonstrated. Guesses and behavioral "trials" are seldom completely random, and the responses are seldom optimally selected with respect to reinforcement contingencies; to some extent, the same analogies could apply to all levels of evolutionary processes. Previous bias can play a significant, constraining role in determining which changes are more or less likely. Learning, then, adds an additional layer of evolution-like processes to the hierarchy, and producing this ability is one of the major reasons that brains evolved. *But what is going on in the brain to support this process?*

A number of scientists and philosophers have turned to Darwinian processes for clues to try to explain the middle level of brain functions. The biologist Gerald Edelman (1987), the neuropsychologist Michael Gazzaniga (1985), the neurologist Jason Brown, the physiologist William Calvin (1987; 1989), the philosopher of mind Daniel Dennett (1991), and the evolutionary theorist Henry Plotkin are just a few prominent thinkers who proposed variants on Darwinian theories of learning processes.

In general, some approach along these lines must be appropriate. No attempt has been made here to survey the similarities and differences, or strengths and weaknesses of these various theories, and there are many important differences that will require empirical modeling and testing to ultimately sort out. Rather, let's end with some general comments on how the underlying Darwinian processes should help researchers think about alternative models, for these models, too, can be stopped short of a thorough-going Darwinian analysis the same way that the standard "design by natural selection" models have: by insinuating design logic somewhere in the middle without noticing it.

Patterns of neural activity determine which neural circuits survive and which are eliminated in development. The sculpting of the connectional architecture of the brain during its development is thus a direct reflection of the underlying logic of neural information processing. This process is a Darwinian process. Thus, one may gain some hints by looking at what selectively favors the persistence of certain patterns of connection to others during development.

As noted above, the answer appears to be the degree of coherence of the signals carried by competing axons as they converge at each neuron. Dyssynchrony is selected against, and axons transmitting signals that are least like the collective mean pattern tend to be eliminated. Each neuron is essentially tallying votes and rewarding axons who voted for the winning pattern. This process offers two important clues concerning the global logic of neural signal processing: (1) correlated temporal patterning is likely to be important in selecting which signals are most likely to persist and spread through the nervous system and (2) different temporal patterns of activity are likely to be in competition with one another for representation in succeeding moments and for being passed to other sites within the brain.

The likely importance of correlated temporal patterning is not new and is based largely on the pioneering work of Donald Hebb (Hebb, 1949). Since the discovery of electrical activity emanating from the cerebral cortex at characteristic frequencies (as measured by **electroencephalography**), neuroscientists have felt that some aspect of cyclic signal processing might be critical to organizing global brain functions. Gerald Edelman (1987) and Francis Crick (Crick and Koch, 1990) are two prominent nobel laureates who have suggested alternative theories in which the establishment of temporal synchrony between widely separated brain structures is the key step in recruiting them to contribute to a common cognitive task.

The production of an integrated behavioral response inevitably requires the concerted participation of a wide range of brain structures, and there is considerable evidence that alternative response tendencies arise in different brain regions and compete to recruit correlated activity in other brain regions. Ultimately, the patterns that have recruited the most correlated activity and that have displaced the alternatives control behavior. This process seems particularly evident in the motor system in which the coordinated activity of large sectors of motor cortex are necessary to initiate directed limb movement.

But where do such patterns originate in the first place? Are there specific neurons that are the drivers? Probably the key to the "prime mover" problem can be found in the intrinsic noisiness of neural signal production. Neurons are probably unreliable signal transducers and, because of intrinsic metabolic and molecular fluctuations, are spontaneously uncorrelatedly active. They are a source of intrinsic noise. Rather than being a problem, however, spontaneous noisiness may be the source of pattern generation — or, rather, that from which neural selection processes can extract adaptive patterns. The selection biases are produced by extrinsic signals supplied by the senses, intrinsic biases supplied by homeostatic mechanisms, and the architecture of connectivity, as determined by development and by the way memory traces bias the connections between patterns. There need be no prime mover in evolutionary processes, no specific instigator of a given response, and no localized organizing initiating center. Only sources of variety and biases that select from it are necessary.

Though this account of the global logic of neural information processing is highly speculative and vague (as all current theories of global brain function must be), it offers a model for understanding brain functions that is consistent with the overall dynamic of organism evolution and development in ways that approaches from design are not. Though we may be able to develop algorithmic and deductive models that capture some superficial aspects of human cognitive processes, they are unlikely to have any deep similarity to what really goes on inside brains, since this is inevitably statistical and Darwinian in character at almost every level.

Thus, evolution is the origin of intelligence in both senses. During a period of billions of years, life has evolved to be more capable of internally representing ever more complex, multileveled environmental relationships and producing adaptive responses to the predicaments they pose. The process has been spontaneous, evolving means of embedding evolutionary processes within other evolutionary processes to produce complex multilevel organisms with greatly amplified adaptive information generating capabilities.

The most serious barrier to developing a full explanation of the evolution of intelligence, complex behavioral abilities, and mechanisms underlying conscious intentionality has probably been the failure to understand organism function from a fully Darwinian perspective. *We have been ensnared by the reasonableness of engineering and design metaphors,* and have used them as short cuts to avoid a full evolutionary accounting of the many levels of ontogenetic, physiological, and social mechanisms involved in the generation of adaptive complexity. Following the design metaphor, we have been tempted to think of brain evolution as the addition of modules, of brain development as design from a genetic plan, and of cognition as a collection of algorithmic processes.

The key insight that leads beyond these theoretical cul-de-sacs is the realization that the evolutionary processes that have produced our bodies and minds operate at *all* levels. Though genetic information provides the outside constraints within which all other organismic information processes must be encapsulated, it accounts for only a very minute fraction of the total information that must be generated to build an animal body and an even smaller fraction of what must be generated to specify the ongoing behavioral adaptations that an animal produces during its lifetime. This information must be spontaneously generated during development and must be generated moment-by-moment in behavior and cognition. The sources of new information are distributed in invariant relationships that already exist and are internalized by higher-order Darwinian processes.

Some of the most compelling analyses of high-level cognitive processes have been presented in terms of functional design, especially those thought to be strongly dependent on unlearned information. A well-known example is the cognitive basis for grammatical analysis and language acquisition characterized by theories of an innate universal grammar (for a current interpretation of this position see Chapter 6 and Pinker, 1994). Currently, the wonderfully suggestive title *Darwinian Psychology* has become synonymous with a whole class of similar accounts of behavioral and cognitive predispositions (see Chapter 4).

In summary, these views share in common the notion that the human mind can be modeled as a large collection of algorithm-like information processors or computational modules that have been designed by evolution to function in highly selective information processing domains, which essentially carry out their operations in isolation from other modules with which they share inputs and outputs. This is an explicit view of intelligence as designed, albeit in piecemeal fashion, with natural selection as the designer.

Though the Darwinian physiological approach outlined here does not exclude the possibility that highly modular sensorimotor or cognitive predispositions may be present from birth, it does suggest that modeling the predispositions in terms of **algorithms**, deductive rule systems, or lists of outputs provides misleading **heuristics** for understanding the evolutionary processes and information processes that underlie them. Moreover, the Darwinian perspective developed here suggests that such a modular design strategy is the exception and not the rule; that is, if there is an efficient way of achieving the same adaptive end without the need for previous specification of the information, the efficient way will be used. The molecular and cellular processes that produce brains operate according to a logic very unlike that of a computer programmer devising a series of operations to serve a specified purpose, and what these processes produce is a result very unlike a computer algorithm. Even if one argues that a given neural operation is strongly prespecified by intense selection, the challenge is to discover how the design metaphors used to describe its operation can be realized in terms of a Darwinian ontogenetic process. It may be better to recast the models in Darwinian terms from the start and abandon the design metaphor entirely.

In general, the evolution of behavioral information processing has not produced brains that are analogues of general-purpose neural computers, nor has it produced brains that are merely collections of special-purpose minicomputers, each designed to fit a particular evolutionary task. Though it has produced approximations to both these design strategies in some of its different parts, these are extreme special cases in neural development (at least in mammal brains, though not necessarily in small invertebrate brains). For the most part, what is behind the intelligent design and function of brains — even in these special cases — is the evolution of subsidiary Darwinian processes, embedded one in the other, that are thereby able to vastly amplify the scope of the adaptive search process and relive the genome of this impossible task. Brain functions are the most elaborated expressions of such a multilevel hierarchy of Darwinian processes.

THE EXPERIENCE OF INTELLIGENCE

What insights does the Darwinian approach to organism design have to offer in approaching that last terra incognita of neuroscience: consciousness and the experience of generating intelligent thought and action?

Philosophers have long struggled with the problem of making sense of the subjective experience of consciousness in the context of the deterministic laws of the chemistry and physics that must underlie mental processes. We all have an undeniable experience of being sources of action and consideration, of deciding what to do, of resisting compulsion, and of reasoning about the best course of action before we act accordingly. We do not experience life as one might imagine that a deterministic robot does, for which present activities are entirely prefigured in previous states of its mechanism or are dictated by the execution of the next line in a computer program code. The life of the mind, so to speak, does not have a mechanical character. Indeed, we easily distinguish times when we act on impulse or unconsciously perform a rote response from times when we act intelligently and rationally — or so we think.

Nevertheless, except for processes at the subatomic level (where a few desperate theorists have looked for a loophole), physical events are remarkably well determined by previous conditions, which is the source of the age-old conundrum about whether humans have free will or are blindly determined in behavior — a problem that more than once has found its way into courtroom questions of sanity and responsibility. If consciousness is a function of the actions of matter and energy at the macroscopic level, how could humans really be the spontaneous originators of *new* actions and thoughts, not actions or thoughts prefigured in what came before? Are not humans mere mechanisms, mere chemical machines performing determined computations?

Not all mechanisms are mechanisms in the way we tend to think of clockworks: boringly mechanical and predictable. Specifically, organisms are not just somewhat messy and unpredictable on the surface; they are deeply indeterminate systems through and through. In this regard, they are the opposite of clockworks. How one behaves moment-to-moment is the result of one level of unspecified, statistical evolution-like process embedded within another, embedded within another, and so on. The outside limiting level of this process is called phylogenetic evolution of species, and the most embedded level, conditioned and functionally removed from simple determinative relationships by all the other levels in between, is called mental experience and action.

What exactly is the form of the experience of being an intelligent actor in the world? It is the experience of originating, moment-by-moment, new, detailed, appropriate representations of an ever-changing environment and spontaneously generating adaptive responses to that environment without any detailed, predetermined compulsion. We have considered analogues to this problem before. The problem of spontaneously generating new, complex adaptive responses where none previously existed, of generating useful information from scratch to fit in with and take advantage of a new environment even as it presents itself, is another version of the adaptive pattern generating problems at all levels of organism function.

Of all the problems of generating adaptively complex patterns that we have considered, the problem of generating intelligent behavior is one of the most demanding. Faced with the onslaught of an unimaginably diverse and unpredictable environment, with new details flooding in moment by moment and demanding an adaptive response, only an evolution-like process could be capable of generating *online* a sufficiently flexible variety of appropriate responses.

There is, as we have seen, only one class of processes that are autonomously able to generate new adaptive complexity: *Darwinian selection processes.* Organisms have evolved to take advantage of this means of generating adaptive information at many levels of functioning to overcome the immense shortfall of structural information that can be provided within a genome and maintained by natural selection and also because it offers a source of highly robust regulatory flexibility. Intelligence is merely the top level of the hierarchy. Our brains keep up with the massively information-rich, input-output context presented to them by continuously generating new adaptive complexity. In a parallel to the generation of the complex patterns of connectivity that constitute our brains, the flow of signals within them carries on the ubiquitous, evolutionary dynamic, spontaneously generating an immense variety of patterns that are sculpted and evolved in microcosm under selection pressures provided by the senses and past memories, to a level of complexity that matches the complexity presented to it. Intelligence *is* a species of evolution. Moreover, the experience of being an intelligent organism *is* the experience of being an evolutionary process in action. It is not just an epiphenomenal illusion that we feel as though we are the originators of new information in the world and that we have an experience of self-determined agency. As evolutionary processes, we are what our experience suggests.

A FINAL NOTE

I offer a detailed alternative theory of language evolution, acquisition, and neural processing based on this Darwinian analysis in my book, *The Idea That Changed the Brain.* Unfortunately, there is not space to expand on these ideas in this short chapter. The general logic of the analysis recognizes yet another important level of evolution-like process as the critical factor at all levels of the brain–language relationship: the **parallel evolution** of representational systems outside the body-symbolic evolutionary processes. Languages, too, are naturally evolved "organisms," not artifacts or inventions. This realization should make it clear that even the use of rule-governed models of grammar and syntax are probably best thought of only as heuristic accounts and may need to be tempered to take this into account. The evolution of languages and the associated symbol systems of cultures have an independent adaptive dynamic, which contributes also to how brains adapt to languages. More specifically, brains and languages have co-evolved, but because languages are far more facile, they have done most of the evolving and adapting to human brains. This is the key to recognizing the Darwinian alternative to theories of innate grammatical knowledge. Languages have evolved to be adapted to the peculiar learning predispositions of very young children, mostly not the other way around.

REFERENCES

Balaban, E., Teillet, M.A., and Le Douarin, N. 1988. Application of the quail-chick chimera system to the study of brain development and behavior. *Science 241*: 1339–1342.

Boncinelli, E., Gulisano, M., and Broccoli, V. 1993. *Emx* and *Otx* homeobox genes in the developing mouse brain. *J. Neurobiol. 24*: 1356–1366.

Boncinelli, E. and Mallamaci, A. 1995. Homeobox genes in vertebrate gastrulation. *Curr. Opin. Gen. Devel. 5*: 619–627.

Calvin, W. 1987. The brain as a Darwin Machine. *Nature 330*: 33–34.

Calvin, W. 1989. A global brain theory. *Science 240*: 1802–1803.

Cowan, W., Fawcett, J., O'Leary, D., and Stanfield, B. 1984. Regressive events in neurogenesis. *Science 255*: 1258–1265.

Crick, F. and Koch, C. 1990. Towards a neurobiological theory of consciousness. *Semin. Neurosci. 2*: 263–275.

Dawkins, R. 1976. *The Selfish Gene* (Oxford: Oxford Univ. Press), 224 pp.

Dawkins, R. 1986. *The Blind Watchmaker* (New York: Norton), 332 pp.

Deacon, T. 1990. Rethinking mammalian brain evolution. *Am. Zool. 30*: 629–705.

Deacon, T. 1995. On telling growth from parcellation in brain evolution. In: Alleva, E., Fasolo, A., Lipp, H-P., Nadel, L., and Ricceri, L. (Eds.), *Behavioral Brain Research in Naturalistic and Semi-Naturalistic Settings* (Dordrecht, Netherlands: Kluwer), pp. 37–62.

Deacon, T. In Press. *The Idea That Changed the Brain: The Coevolution of the Human Brain and Language* (New York: Norton).

Deacon, T., Pakzaban, P., Burns, L., Dinsmore, J., and Isacson, O. 1994. Cytoarchitectonic development, axon-glia relationships and long distance axon growth of porcine striatal xenografts in rats. *Exp. Neurol. 130*: 151–167.

Dennett, D. 1991. *Consciousness Explained* (Boston: Little Brown), 511 pp.

Doron, N. and Wollberg, Z. 1994. Cross-modal neuroplasticity in the blind mole rat *Spalax ehrenbergi*: A WGA-HRP tracing study. *Neuroreport 5*: 2697–2701.

Edelman, G. 1987. *Neural Darwinism: The Theory of Neuronal Group Selection* (New York: Basic Books), 371 pp.

Figdor, M. and Stern, C. 1993. Segmental organization of embryonic diencephalon. *Nature 363*: 630–634.

Finlay, B. and Darlington, R. 1995. Linked regularities in the development and evolution of mammalian brains. *Science 268*: 1578–1584.

Finlay, B., Wikler, K., and Sengelaub, D. 1987. Regressive events in brain development and scenarios for vertebrate brain evolution. *Brain Behav. Evol. 30*: 102–117.

Frantz, G., Weimann, J., Levin, M., and McConnell, S. 1994. *Otx1* and *Otx2* define layers and regions in developing cerebral cortex and cerebellum. *J. Neurosci. 14*: 5725–5740.

Frost, D.O. and Metin, C. 1985. Induction of functional retinal projections to the somatosensory system. *Nature 317*: 162.

Gazzaniga, M. 1985. *The Social Brain: Discovering the Networks of the Mind* (New York: Basic Books), 219 pp.

Gould, S.J. and Lewontin, R.C. 1979. The spandrels of San Marco and the Panglossian program: A critique of the adaptationist program. *Proc. R. Soc. London 205*: 281–288.

Hebb, D. 1949. *The Organization of Behavior: A Neuropsychological Theory* (New York: Wiley), 335 pp.

Hoey, T. and Levine, M. 1988. Divergent homeobox proteins recognize similar DNA sequences in *Drosophila*. *Nature 332*: 858–861.

Holland, P., Ingham, P., and Krauss, S. 1992. Development and evolution. Mice and flies head to head. *Nature 358*: 627–628.

Isacson, O., Deacon, T., Pakzaban, P., Galpern, W., Dinsmore, J., and Burns, L. 1995. Transplanted xenogeneic neural cells in neurodegenerative disease models exhibit remarkable axonal target specificity and distinct growth patterns of glial and axonal fibres. *Nat. Med. 1*: 1189–1194.

Isacson, O. and Deacon, T. In Press. Presence and specificity of axon guidance cues in the adult brain: Evidence from xenografts. *Neuroscience.*

Jackendoff, R. 1992. *Languages of the Mind* (Cambridge: MIT Press), 200 pp.

Kappen, C. and Ruddle, F. 1993. Evolution of a regulatory gene family: HOM/HOX genes. *Curr. Opin. Gen. Devel. 3*: 931–938.

Katz, J. and Lasek, R. 1983. Evolution of the nervous system: Role of ontogenetic mechanisms in the evolution of matching populations. *Proc. Natl. Acad. Sci. USA 75*: 1349–1352.

Kauffman, S. 1992. *Origins of Order: Self-Organization and Selection in Evolution* (Oxford: Oxford Univ. Press), 709 pp.

Kessel, M. 1993. Reversal of axonal pathways from rhombomere 3 correlates with extra Hox expression domains. *Neuron 10*: 379–393.

Keynes, R. and Krumlauf, R. 1994. Hox genes and regionalization of the nervous system. *Ann. Rev. Neurosci. 17*: 109–132.

Killackey, H., Chiaia, N., Bennett-Clarke, C., Eck, M., and Rhoades, R. 1994. Peripheral influences on the size and organization of somatotopic representations in the fetal rat cortex. *J. Neurosci. 14*: 1496–1506.

Law, M. and Constantine-Paton, M. 1981. Anatomy and physiology of experimentally induced striped tecta. *J. Neurosci. 1*: 741–759.

McGinnis, W. and Kuziora, M. 1994. The molecular architects of body design. *Sci. Am. 270*: 58–66.

Molnçr, Z. and Blakemore, C. 1991. Lack of regional specificity for connections formed between thalamus and cortex in coculture. *Nature 351*: 475–477.

O'Leary, D. and Stanfield, B. 1989. Selective elimination of axons extended by developing cortical neurons is dependent on regional locale experiments utilizing fetal cortical transplants. *J. Neurosci. 9*: 2230–2246.

O'Leary, D. 1992. Development of connectional diversity and specificity in the mammalian brain by the pruning of collateral projections. *Curr. Opin. Neurobiol. 2*: 70–77.

Paley, W. 1802. *Natural Theology: Or, Evidences of the Existence and Attributes of the Deity, Collected from the Appearances of Nature* (London: E. Paulder), 586 pp.

Pinker, S. 1994. *The Language Instinct: How the Mind Creates Language* (New York: William Morrow), 494 pp.

Purves, D. 1988. *Body and Brain. A Trophic Theory of Neural Connections* (Cambridge: Harvard Univ. Press), 231 pp.

Purves, D. and Lichtman, J. 1980. Elimination of synapses in the developing nervous system. *Science 210*: 153–157.

Roth, G., Nishikawa, K.C., Naujoks-Manteuffel, C., Schmidt, A., and Wake, D.B. 1993. Paedomorphosis and simplification in the nervous system of salamanders. *Brain Behav. Evol. 42*: 137–170.

Rubenstein, J., Martinez, S., Shinmamura, K., and Puelles, L. 1994. The embryonic vertebrate forebrain: The prosomeric model. *Science 266*: 578–580.

Sacher, G. and Staffeldt, E. 1974. Relation of gestation time to brainweight for placental mammals: Implications for the theory of vertebrate growth. *Am. Nat. 108*: 593–615.

Simeone, A., Acampora, D., Gulisano, M., Stornaiuolo, A., and Boncinelli, E. 1992. Nested expression domains of four homeobox genes in developing rostral brain. *Nature 358*: 687–690.

Stanfield, B. and O'Leary, D. 1985. Fetal occipital cortical neurons transplanted to the rostral cortex can extend and maintain a pyramidal tract axon. *Nature 313*: 135–137.

Sur, M., Garraghty, P., and Roe, A. 1988. Experimentally induced visual projections into auditory thalamus and cortex. *Science 242*: 1437–1441.

Wilczynski, W. 1984. Central neural systems subserving a homoplasous periphery. *Am. Zool. 24*: 755–763.

Yamamoto, N., Yamada, K., Kurotani, K., and Toyama, K. 1992. Laminar specificity of extrinsic cortical connections studied in coculture preparations. *Neuron 9*: 217–288.

CHAPTER 6

EVOLUTIONARY BIOLOGY AND THE EVOLUTION OF LANGUAGE

Steven Pinker*

IN BIOLOGY UNIQUENESS IS COMMON

The elephant's trunk is 6 feet long, 1 foot thick, and contains 60,000 muscles. Elephants can use their trunks to uproot trees, stack timber, or carefully place huge logs into position when recruited to build bridges. They can curl the trunk around a pencil and draw characters on letter-size paper. With the two muscular extensions at the tip of the trunk, they can remove a thorn; pick up a pin or a dime; uncork a bottle; slide the bolt off a cage door and hide it on a ledge; or grip a cup, without breaking it, so firmly that only another elephant can pull it away. The tip is sensitive enough for a blindfolded elephant to ascertain the shape and texture of objects. In the wild, elephants use their trunks to pull up clumps of grass and tap them against their knees to knock off the dirt, to shake coconuts out of palm trees, and to powder their bodies with dust. They use their trunks to probe the ground as they walk, avoiding pit-traps, and to dig wells and siphon water from them. Elephants can walk underwater on the beds of deep rivers or swim like submarines for miles, using their trunks as snorkels. They communicate through their trunks by trumpeting, humming, roaring, piping, purring, rumbling, and making a crumpling-metal sound by rapping the trunk against the ground. The trunk is lined with chemoreceptors that allow the elephant to smell a python hidden in the grass or food a mile away (Williams, 1989; Carrington, 1958).

Elephants are the only living animals that possess this extraordinary organ. Their closest living relative is the hyrax, a mammal that you would probably not be able to tell from a large guinea pig. Until now you have probably not given the uniqueness of the elephant's trunk a moment's thought. Certainly no biologist has made a fuss about

*Department of Brain and Cognitive Sciences, Massachusetts Institute of Technology, Cambridge, MA 02139 USA

it. But now imagine what might happen if biologists were elephants. Obsessed with the unique place of the trunk in nature, they might ask how it could have evolved, given that no other organism has a trunk or anything like it. One school might try to think up ways to narrow the gap. They would first point out that the elephant and hyrax share at least 90% of their DNA, so they could not be all that different. They might say that the trunk must not be as complex as everyone thought; perhaps the number of muscles had been miscounted. They might further note that the hyrax really *does* have a trunk, but it has somehow been overlooked; after all, hyraxes do have nostrils. Though their attempts to train hyraxes to pick up objects with their nostrils would have failed, some in this school might trumpet their success at training the hyraxes to push toothpicks around with their tongues, noting that stacking tree trunks or drawing on pieces of paper differ from it only in degree.

The opposite school, maintaining the uniqueness of the trunk, might insist that it appeared all at once in the offspring of a particular trunkless elephant ancestor, the product of a single dramatic mutation. Or they might say that the trunk somehow arose as an automatic byproduct of the elephant's having evolved a large head. They might add another paradox for trunk evolution: The trunk is absurdly more intricate and well-coordinated than any ancestral elephant would have needed.

These arguments might strike us as peculiar, but every one of them has been made by scientists of a different species about a complex organ that that species alone possesses — language. Noam Chomsky (a famous scientist in this field) and some of his fiercest opponents agree on one thing, namely, that a uniquely human language instinct seems to be incompatible with the modern Darwinian theory of evolution, in which complex biological systems arise by the gradual accumulation over generations of random genetic mutations that enhance reproductive success. Either there is no specific human language instinct, or it must have evolved by other means. Because I have argued extensively that there is one (Pinker, 1994), but would certainly forgive anyone who would rather believe Darwin than believe me, I would like to show that one does not have to make such a choice. Though we know few details about how language evolved, there is no reason to doubt that the principal explanation is the same as for any other complex instinct or organ, Darwinian natural selection. (See also Pinker and Bloom, 1989; Pinker, in press; Hurford, 1989, 1991; Newmeyer, 1991; Brandon and Hornstein, 1986; Corballis, 1991.)

COULD LANGUAGE BE UNIQUE TO HUMANS?

Language is obviously as different from other animals' communication systems as the elephant's trunk is different from other animals' nostrils. Nonhuman communication systems are based on one of three designs: (1) a finite repertory of calls (one for warnings of predators, one for claims to territory, and so on), (2) a continuous analog signal that registers the magnitude of some state (for example, the livelier the dance of the bee, the richer the food source that it is telling its hivemates about), or (3) a series of random variations on a theme (a birdsong repeated with a new twist each time: Charlie Parker with feathers) (Wilson, 1972; Gould and Marler, 1987). Human language, of course, has a very different design. The discrete combinatorial system called **grammar** makes human language essentially *infinite* (that is, there is no limit to the number of complex words or sentences in a language); *digital* (its infinity achieved by rearranging discrete elements in particular orders and combinations, not by varying some signal along a continuum like the mercury in a thermometer); and *compositional* (that is, each infinite combination has a different meaning predictable from the meanings of its parts and the rules and principles arranging them).

Even the seat of human language in the brain is special. The vocal calls of primates are controlled not by the cerebral cortex portion of the brain but by the phylogenetically older and **subcortical** neural systems in the brainstem and limbic system that are heavily involved in emotion. Human vocalizations other than language, such as sobbing, laughing, moaning, and shouting in pain, are also controlled subcortically. Language itself, of course, is seated in the **cerebral cortex**, primarily in what are known as the left perisylvian regions (Deacon, 1988, 1989; Caplan, 1987; Myers, 1976; Robinson, 1976).

Do Chimpanzees Have Language?

Some psychologists believe that changes in the vocal organs and in the neural circuitry that produces and perceives speech sounds are the *only* aspects of language that evolved in our species. According to this view, there are a few general learning abilities found throughout the animal kingdom, and these few work most efficiently in humans. At some point in history, language was invented and refined, and we have been learning it ever since. Accordingly, chimpanzees are viewed as the second-best learners in the animal kingdom; they should be able to acquire a language, too, albeit a simpler one. All it takes is a teacher, and a "sensorimotor channel" that chimpanzees can control. Indeed, beginning in the late 1960s, several famous projects (Wallman, 1992) claimed to have taught language to chimpanzees; for example, to the chimpanzees Sara (Premack and Premack, 1972; Premack, 1985), Kanzi (Savage-Rumbaugh, 1991), and Washoe (Gardner and Gardner, 1969).

These claims have not only captured the public's imagination (having been played up in many popular science books, magazines, and television programs), but have also captivated many scientists who see the projects as a healthy deflation of our species' arrogant chauvinism (Sagan and Druyan, 1992). Some psycholinguists — recalling Darwin's insistence on the gradualness of evolutionary change — seem to believe, moreover, that a detailed examination of chimps' behavior is unnecessary: They *must* have language, as a matter of principle. For example, Elizabeth Bates, Donna Thal, and Virginia Marchman (1991, pp. 30, 35) wrote:

> If the basic structural principles of language cannot be learned (bottom up) or derived (top down), there are only two possible explanations for their existence: either Universal Grammar was endowed to us directly by the Creator, or else our species has undergone a mutation of unprecedented magnitude, a cognitive equivalent of the Big Bang. . . . We have to abandon any strong version of the discontinuity claim that has characterized generative grammar for thirty years. We have to find some way to ground symbols and syntax in the mental material that we share with other species.

When the chimps' abilities are looked at on their own, though, free from theoretical and ethical preconceptions, different conclusions emerge (Terrace et al., 1979; Seidenberg and Petitto, 1979, 1987; Petitto and Seidenberg, 1979; Wallman, 1992). For one thing, the apes most definitely have *not* "learned" **American Sign Language** (ASL), as has been frequently announced. This preposterous claim is based on the myth that ASL is a crude system of pantomimes and gestures, rather than a full language with complex phonology, morphology, and syntax. Furthermore, the chimps' grammatical abilities are close to nil. Whereas typical sentences from a 2-year-old child are "Look at that train Ursula brought" and "We going turn light on so you can't see," typical sentences from a language-trained chimp are "Me banana you banana me you give" and "Give orange me give eat orange me eat orange give me eat orange give me you."

As discussed in Chapter 3, Sue Savage-Rumbaugh and her colleagues have claimed that even though common chimps cannot be taught sign language, pygmy chimps (bonobos) can be taught a language that uses visual lexigrams. (Why pygmy chimps should be expected to do so much better than members of their sibling species is not clear; contrary to often repeated suggestions, pygmy chimps are no more closely related to humans than common chimps are.) A crowning linguistic achievement of Kanzi, a bonobo chimp, was said to be a kind of three-symbol "sentence," but the so-called sentences are all simple chains such as the symbol for "chase," followed by the symbol for "hide," followed by pointing to a person. Kanzi wants to do the chasing and hiding. That is, Kanzi appears to be using fixed formulas, with no internal structure, and the constructions are not even three symbols long.

Putting aside the empirical question of what chimpanzees in fact have accomplished, let us return to the theoretical question of whether evolutionary theory requires that chimpanzees "must" have acquired some version of language. If human language is unique in the modern animal kingdom, as it appears to be, the implications for a Darwinian account of its evolution would be as follows: none. A language instinct unique to modern humans poses no more of a paradox than does a trunk unique to modern elephants — no contradiction, no Creator, no "Big Bang."

Modern evolutionary theorists are alternately amused and annoyed by a curious fact. Though most educated people profess to believe in Darwin's theory, what they really believe is a slightly modified version of the ancient theological notion of the **Great Chain of Being** — the idea that all species are arrayed in a linear hierarchy with humans at the top. Darwin's contribution, according to this belief, showed that each species on the ladder evolved from the species one rung down instead of being allotted its rung by God. Dimly remembering their high school biology classes that took them on a tour of the phyla from primitive to modern, people think roughly of the pathway shown in Figure 6.1: Amoebas begat sponges which begat jellyfish which begat flatworms which begat trout which begat frogs which begat lizards which begat dinosaurs which begat anteaters which begat monkeys which begat chimpanzees which begat us. (A few steps have been skipped for the sake of brevity.)

Bushes and Ladders — The Phylogeny of Language

Hence, the paradox: Humans enjoy language, but their neighbors on the immediately adjacent rungs experience nothing of the kind. We expect a fade-in, but we see a Big Bang. But evolution did not make a ladder; it made a *bush* (Mayr, 1982; Dawkins, 1986; Gould, 1985). We did *not* evolve from chimpanzees. We and chimpanzees evolved from a common ancestor, now long extinct. The human-chimp ancestor did not evolve from monkeys, but from an even older ancestor of both, also (even longer) extinct. And so on, back to our single-celled forebears. Paleontologists often say that to a first approxima-

The Wrong Theory

Amoebas
↓
Sponges
↓
Jellyfish
↓
Flatworms
↓
Trout
↓
Frogs
↓
Lizards
↓
Dinosaurs
↓
Anteaters
↓
Monkeys
↓
Chimpanzees
↓
Homo sapiens

FIGURE 6.1

An abbreviated version of the Great Chain of Being in which all species are arrayed in a ladder-like hierarchy from the most primitive (represented here by amoebas) to *Homo sapiens*, the most advanced. This ancient theological notion is wrong — evolution did not make a ladder, it made a *bushy tree* like that shown in Figure 6.2.

tion, all species are extinct (99% or more is the usual estimate). The organisms we see around us are distant cousins, not great-grandparents. They are a few scattered twig-tips of an enormous, bush-like tree whose branches and trunk are no longer with us. Figure 6.2, though greatly simplified, summarizes the relationships.

Zooming in on our branch of the bushy tree (Figure 6.3), we see that chimpanzees are off on a separate subbranch. They sit neither beneath nor above us. We also see that a form of language could first have emerged at the position of the arrow in Figure 6.3, after the branch leading to humans split off from the one leading to chimpanzees. The result would be languageless chimps and approximately 5 to 7 million years in which language could have gradually evolved. Indeed, we should zoom in even closer, because separate species do not mate and produce baby species; organisms of the same species mate and produce baby organisms. "Species" is an abbreviation for a chunk of a vast family tree composed of individuals, such as the *particular* gorilla, chimp, australopithecine, *Homo erectus,* archaic *H. sapiens*, Neanderthal, and modern *H. sapiens* shown in the family tree in Figure 6.4.

If the first trace of protolanguage-ability appeared in the ancestor at the arrows in Figures 6.3 and 6.4, there could have been about 350,000 generations between then and now for the ability to have been elaborated and fine-tuned to the **Universal Grammar** we see today. For all we know, language could have had a gradual fade-in, even if no extant species, not even our closest living relatives, the chimpanzees, have it. There were plenty of organisms with intermediate language abilities, but they are all now extinct, long dead.

Here is another way to think about it. People see chimpanzees, the living species closest to us, and are tempted to conclude that at the very least, these closely related cousins must have some ability that is ancestral to language. But because the evolutionary tree is made up of individuals, not species, "the living species closest to us" has no special status. What that species is depends on accidents of extinction. Try the following thought experiment. Imagine that anthropologists discover a relict population of *Homo habilis* in some remote highland. *H. habilis* would now be viewed as our closest living relative. Would that take the pressure off chimps, so it is not so important that they have something like language after all?

Or, do the experiment the other way around. Imagine that some epidemic had wiped out all the apes several thousand years ago. Would Darwin be in danger unless we showed that monkeys had language? If you are inclined to answer yes, push the thought experiment one branch up: Imagine that in the past some extraterrestrials developed a craze for primate fur coats and hunted and trapped all the primates to extinction except hairless humans. Would insectivores such as anteaters have to shoulder the protolanguage burden? What if the extraterrestrials went for mammals in general? Or developed a taste for vertebrate flesh, but spared us because they like the sitcom reruns that we have inadvertently broadcast into space? Would we then have to look for talking starfish? Or have the notion that the syntax of language is grounded in the mental material we share with starfish?

Obviously not. Our brains, and chimpanzee brains, and anteater brains, have whatever wiring they have. Their wiring cannot change, depending on which other species a continent away happen to survive or go extinct. The point of these thought experiments is that the gradualness that Darwin made so much about applies to lineages of individual organisms in a *bushy* family tree, not entire living species in a great chain. For reasons that we will cover soon, an ancestral ape with nothing but hoots and grunts is unlikely to have given birth to a baby who could learn English or the language Kivunjo.

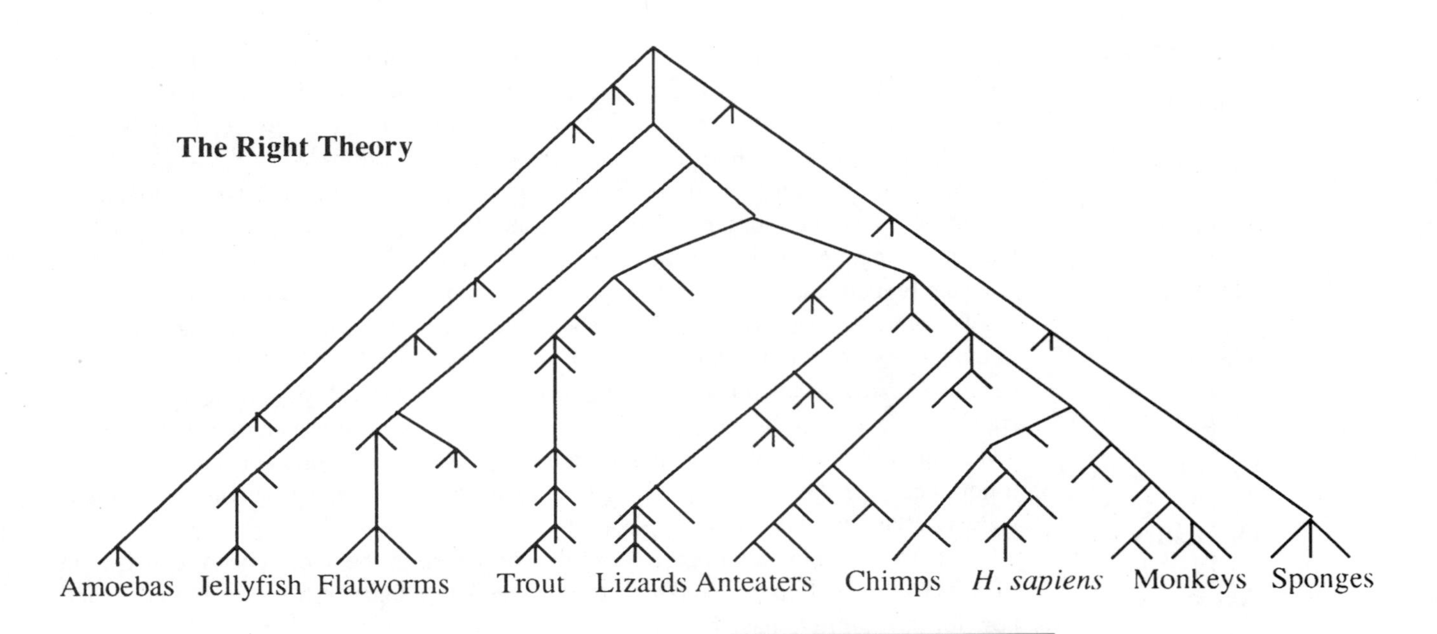

FIGURE 6.2

From bacteria to man, all organisms alive today are distant cousins, twig-tips of an enormous, bush-like Tree of Life. The animal portion of the bushy tree is shown here. Species now exinct, more than 99% of species that have ever lived, make up the branches and trunk of the evolutionary tree; only a few scattered species have survived to the present.

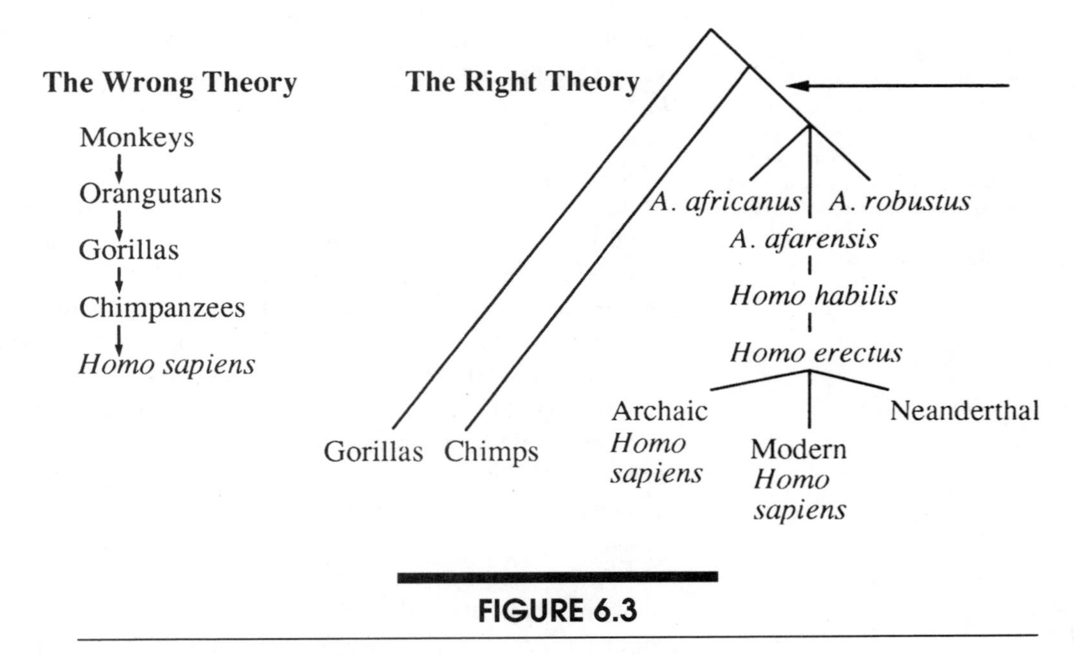

FIGURE 6.3

The primate portion of the bushy Tree of Life ("The Right Theory") compared with the equivalent portion of the Great Chain of Being ("The Wrong Theory"). Chimpanzees are on a subbranch separate from human lineage; chimpanzees are *not* the evolutionary precursors of *Homo sapiens*. Though no one knows for sure, a form of language may first have emerged at the position of the arrow, after the branchlet leading to humans split off from the one leading to chimpanzees.

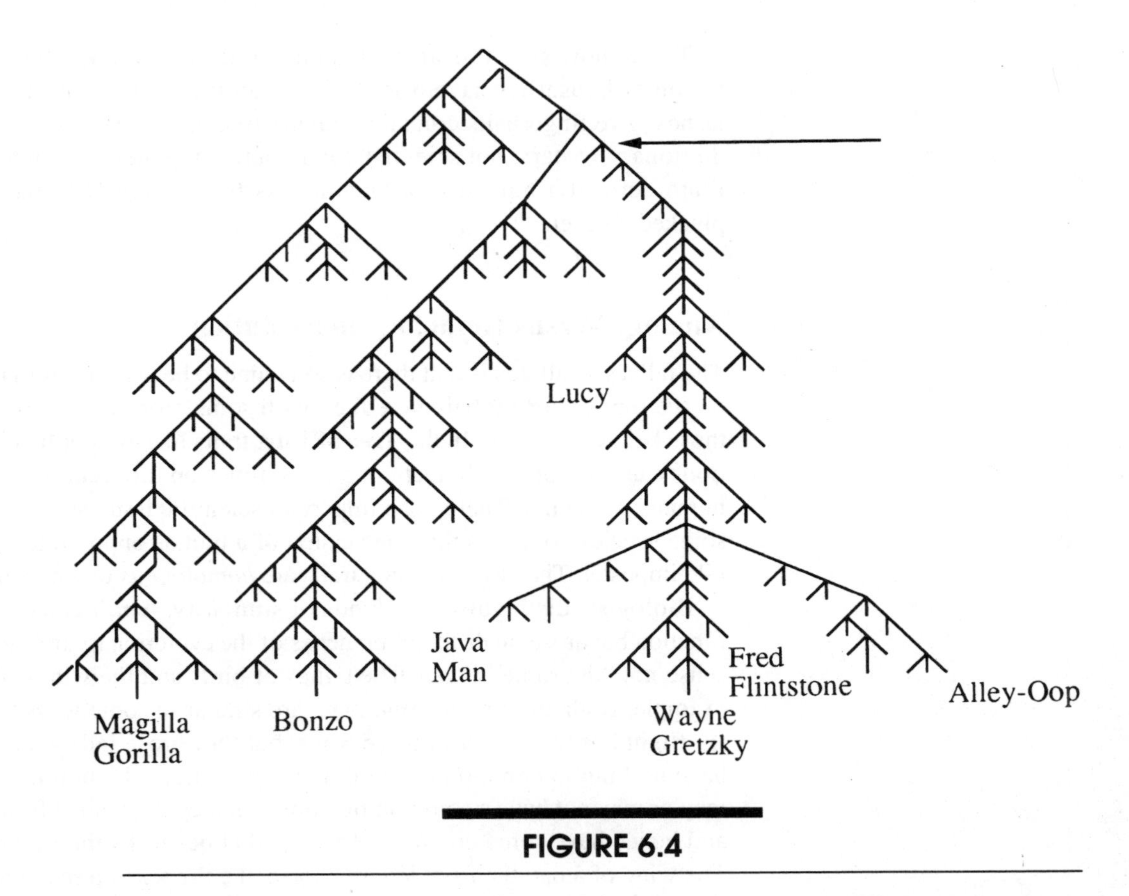

FIGURE 6.4

The bush-like family tree of primates showing the position of a particular gorilla (Magilla), chimp (Bonzo), australopithecine (Lucy), archaic *Homo sapiens* (Java Man), modern *H. sapiens* (Wayne Gretzky), Neanderthal (Fred Flintstone), and *H. erectus* (Alley-Oop). If the first trace of a protolanguage appeared in the ancestor denoted by the arrow, hundreds of thousands of generations would have been available between then and now for elaboration and fine-tuning of the Universal Grammar of today.

But it did not have to; there was a chain of several hundred thousand generations of grandchildren in which such abilities could gradually blossom.

To determine when in fact language began, we have to look at people, look at animals, and note what we see. We cannot use the idea of phyletic continuity to legislate the answer from an armchair.

The difference between bush and ladder also allows us to put a lid on a fruitless and boring debate over what qualifies as True Language. One side, the "Human Uniqueness Team," lists qualities that human language has but that no animal has yet demonstrated: use of symbols displaced in time and space from the items to which they are referring ("referents"), creativity, categorical speech perception, consistent ordering, hierarchical structure, infinity, recursion, and so on. The other side finds a counterexample in the animal kingdom (perhaps budgies can discriminate speech sounds, or dolphins or parrots can attend to word order when carrying out commands, or some songbird can improvise indefinitely without repeating itself) and gloats that the citadel of human uniqueness has been breached. Then, in the face of this criticism, the Human Uniqueness Team relinquishes its uniqueness criterion but emphasizes others or adds new ones to the list, thus provoking angry objections that they are moving the goalposts.

To see how silly this all is, imagine a debate over whether flatworms have true vision or houseflies have true hands. Is an iris critical for vision? What about eyelashes? Are fingernails crucial to having true hands? Who cares? This is a debate for dictionary writers, not scientists. Plato and Diogenes were not doing biology when Plato defined a human as a "featherless biped" and Diogenes refuted him with a plucked chicken.

Analogy Versus Homology in Evolution

The fallacy in all this is that there is some line to be drawn across the ladder, the species on the rungs above it being credited with some glorious trait of ours or its precursor, those below lacking it. In the **Tree of Life**, traits like eyes or hands or infinite vocalizations can arise on any branch, or several times on different branches, some leading to humans, some not. There is an important scientific issue at stake, but it is not whether some species possesses the true version of a trait as opposed to some pale imitation or vile imposter. The issue is which traits are *homologous* to which other ones.

Biologists distinguish two kinds of similarity. **Analogous traits** have a common function but arose on different branches of the evolutionary tree and are in an important sense not "the same" organ. The wings of birds and the wings of bees are a textbook example. Both are used for flight and are similar in some ways because anything used for flight has to be built in those ways, but they arose independently in evolution and have nothing in common beyond their use in flight. **Homologous traits**, in contrast, may or may not have a common function, but they descended from a common ancestor, and hence have some common structure that bespeaks their being "the same" organ. The wing of a bat, the front leg of a horse, the flipper of a seal, the claw of a mole, and the hand of a human all have different functions, but they are all modifications of the forelimb of the ancestor of all mammals. As a result, all share nonfunctional traits such as the number of bones and the way they are connected. To distinguish analogy from homology, biologists usually look at the overall architecture or plan of the organs and focus on their most useless properties, because the useful ones could have arisen independently in two lineages *because* they are useful, a nuisance to taxonomists called **convergent evolution**. We deduce that bat wings are really hands because we can see the wrist and count the fingers and because that is not the only way that nature could have built a wing.

The interesting question is whether human language is homologous, that is, biologically "the same thing," to anything in the modern animal kingdom. Discovering some similarity, such as sequential ordering, is pointless, especially when it is found on a remote branch, such as a bird species, which is surely not ancestral to humans. Here primates are relevant, but the ape trainers and their fans are playing by the wrong rules. Imagine that their wildest dreams are realized and some chimpanzee can be taught to produce real signs, to group and order them consistently to convey meaning, to use them spontaneously to describe events, and so on. Does that show that the human ability to learn language evolved from the chimp ability to learn the artificial sign system? Of course not, any more than seagulls evolved from mosquitos. The symbol system that the chimp learned would have been useful to him and, indeed, the structure was provided by the human inventors.

To check for homology with human language, one must find some *inherent* property (not one taught by experimenters) that reliably emerges in ape sign systems and in human language but that could just as easily have been otherwise. One could look at development, checking for some echo of the standard sequence from syllable babbling to jargon babbling to first words to two-word sequences to a grammar explosion. One could look at developed grammar to see whether apes develop or favor some version of nouns and verbs, inflections, specific types of syntax (for example, X-bar), roots and stems, or auxiliaries in second position inverting to form questions. (These structures are not so abstract as to be undetectable; they leapt from the data when linguists first compared American Sign Language and the language of creoles, for example.) One could also look at neuroanatomy, checking for control by the left perisylvian regions of the cortex, with grammar more anterior, dictionary more posterior. This line of questioning, routine in biology since the nineteenth century, has never been applied to chimp signing, though one can make a good prediction of what the answers would be.

CAN NEW MODULES EVOLVE?

How plausible is it that the ancestor to language first appeared after the branch leading to humans split off from that leading to chimps? Not very, says Philip Lieberman (1990), who believes that vocal-tract anatomy and speech control, not a grammar module, are the only things that were modified in evolution: "Since Darwinian natural selection involves small incremental steps that enhance the present function of the specialized module, the evolution of a 'new' module is logically impossible" (Lieberman, 1990, pp. 741–742). Now, something has gone seriously awry in this logic. Humans evolved from single-celled ancestors. Single-celled ancestors had no arms, legs, heart, eyes, liver, and so on. Therefore, eyes and livers are logically impossible.

The fallacy is that although natural selection involves incremental steps that enhance functioning, the enhancements do not have to be to an existing module. They can slowly build a module from a previously nondescript stretch of anatomy or from the nooks and crannies between existing modules, which the biologists Stephen Jay Gould and Richard Lewontin call "spandrels," from the architectural term.

An example of a brand new module is the eye, which has arisen de novo some 40 separate times in animal evolution. It can begin in an eyeless organism with a patch of skin whose cells are sensitive to light. The patch can deepen into a pit, cinch up into a sphere with a hole in front, grow a translucent cover over the hole, and so on, each step allowing the owner to detect events a bit better. An example of a module growing out of bits that were not originally a module is the elephant's trunk. It is a brand-new organ, but homologies suggest that it evolved from a fusion of the nostrils and some of the upper lip muscles of the extinct elephant-hyrax common ancestor, followed by radical complications and refinements (Mayr, 1982).

Language could have arisen, and probably did arise, in a similar way: by a revamping of primate brain circuits that originally had no role in vocal communication and by the addition of some brand new ones. Al Galaburda (Galaburda and Panya, 1982) and Terrence Deacon (1988, 1989) have discovered areas in monkey brains that correspond in location, input-output cabling, and cellular composition to human language areas. For example, there are homologues to **Wernicke's** and **Broca's areas** of the human brain, and a structure known as the arcuate fasciculus connecting the two, just as in humans. The regions are not involved in producing the monkeys' calls, nor are they involved in producing their gestures. The monkey seems to use the regions corresponding to Wernicke's area and its neighbors to recognize sound sequences and to discriminate the calls of other monkeys from its own calls. The Broca's homologues are involved in control over the muscles of the face, mouth, tongue, and larynx, and various subregions of these homologues receive inputs from the parts of the brain dedicated to hearing, the sense of touch in the mouth, tongue, and larynx, and areas in which information from all the senses converge.

No one knows exactly why this arrangement is found in monkeys and, presumably, in their common ancestor with humans, but it gave evolution some parts it could tinker with to produce human language circuitry, perhaps exploiting the confluence of vocal, auditory, and other signals there, which would be useful for spoken language.

Brand-new circuits in this general territory could have arisen, too. Neuroscientists charting the cortex with electrodes have occasionally found mutant monkeys that have one extra visual map in their brains compared with brains of standard monkeys. A sequence of genetic changes that duplicate a brain map or circuit, reroute its inputs and outputs, and twiddle and tweak its internal connections could manufacture a genuinely new brain module.

ARE CHIMPS AND HUMANS 99% ALIKE?

Brains can be rewired only if the genes that control the wiring have changed. This brings up another bad argument about why chimp signing must be like human language. The argument is based on the finding that chimpanzees and humans share 98% to 99% of their DNA. The implication is that humans must be 99% similar to chimpanzees.

Geneticists are appalled at such reasoning and take pains to stifle it in the same breath that they report their results. The recipe for the embryological soufflé is so baroque that small genetic changes can have enormous effects on the final product. And a 1% difference is not even so small. In terms of the information content in the DNA, it is *10 megabytes*, big enough for Universal Grammar with lots of room left over for the rest of the instructions on how to turn a chimp into a human. Indeed, a 1% difference in total DNA does not even mean that only 1% of human and chimpanzee genes are different. It could, in theory, mean that 100% of human and chimpanzee genes are different, each by 1%. DNA is a discrete combinatorial code, so a 1% difference in the DNA for a gene can be as significant as a 100% difference, just as changing one bit in every byte, or one letter in every word, can result in a new text that is 100% different, not 10% or 20% different. The reason is that DNA encodes the information used by cells to place particular amino acids at specific places in proteins, and even a single amino acid substitution can change the shape of a protein enough to alter its function completely; this is what happens in many fatal genetic diseases. Data on genetic similarity are useful in figuring out how to connect a family tree (solving such questions, for example, as whether gorillas branched off from a common ancestor of humans and chimps or whether humans branched off from a common ancestor of chimps and gorillas) and perhaps even to date the evolutionary divergences using a molecular clock. But they say nothing about how similar the organisms' brains and bodies are.

DID LANGUAGE EVOLVE BY NATURAL SELECTION?

In a chapter of *On the Origin of Species*, Darwin (1859) painstakingly argued that his theory of natural selection could account for the evolution of instincts and of bodies. If language is like other instincts, presumably it also evolved by natural selection, which is the only successful scientific explanation of the source of complex biological structure.

Chomsky, one might think, would have everything to gain by grounding his controversial theory about a *language organ* in the firm foundation of evolutionary theory, and in some of his writings he has hinted at a connection. But more often (1972), he is skeptical:

> It is perfectly safe to attribute this development [of innate mental structure] to "natural selection," so long as we realize that there is no substance to this assertion, that it amounts to nothing more than a belief that there is some naturalistic explanation for these phenomena. . . . In studying the evolution of mind, we cannot guess to what extent there are physically possible alternatives to, say, transformational generative grammar, for an organism meeting certain other physical conditions characteristic of humans. Conceivably, there are none — or very few — in which case talk about evolution of the language capacity is besides the point (pp. 97–98).

And, again (1988):

> Can the problem [the evolution of language] be addressed today? In fact, little is known about these matters. Evolutionary theory is informative about many things, but it has little to say, as of now, of questions of this nature. The answers may well lie not so much in the theory of natural selection as in molecular biology, in the study of what kinds of physical systems can develop under the conditions of life on earth and why, ultimately because of physical principles. It surely cannot be assumed that every trait is specifically selected. In the case of such systems as language . . . it is not easy even to imagine a course of selection that might have given rise to them (p. 167).

What could this renowned linguist possibly mean? Could there be a language organ that evolved by a process different from the one we have always been told is responsible for all other organs? Many psychologists, impatient with arguments that cannot be fit into a slogan, pounce on such statements and ridicule Chomsky as a cryptocreationist. They are wrong, though I think Chomsky is wrong, too.

The Nature of Natural Selection

To understand the issues, we first must understand the logic of Darwin's theory of natural selection. Evolution and natural selection are not the same thing. **Evolution**, the fact that species change over time because of what Darwin called "descent with modification," was already widely accepted in Darwin's time, but was attributed to now discredited processes such as Lamarck's inheritance of acquired characteristics and an internal urge or striving to evolve in a direction of increasing complexity, culminating in humans. What Darwin and Alfred Wallace discovered and emphasized was a particular *cause* of evolution, natural selection. **Natural selection** applies to any set of entities with the properties of *multiplication*, *variation*, and *heredity*. *Multiplication* means that the entities copy themselves, that the copies are also capable of copying themselves, and so on. **Variation** means that the copying is not perfect; errors crop up from time to time, and these errors may give an entity traits that enable it to copy itself at higher or lower rates relative to other entities. **Heredity** means that a variant trait produced by a copying error reappears in subsequent copies, and thus the trait is perpetuated in the lineage.

Natural selection is the mathematically necessary outcome that any traits that foster superior replication will tend to spread through the population over many generations. As a result, the entities will come to have traits that appear to have been designed for effective replication, including means to this end such as gathering energy and materials from the environment and safeguarding them from competitors. The replicating entities are what we recognize as organisms, and the replication-enhancing traits they accumulated by this process are called **adaptations**. (See also Darwin, 1859; Mayr, 1983; Dawkins, 1986; Tooby and Cosmides, 1990; Maynard Smith, 1984, 1986; Dennett, 1983.)

Is Natural Selection a Circular Theory?

At this point many people feel proud of themselves for spotting what they think is a fatal flaw. "Aha! The theory is circular! All it says is that traits that lead to effective replication lead to effective replication. In other words, natural selection is the *survival of the fittest* and the definition of *the fittest* is *those who survive*." Not!!

The power of the theory of natural selection is that it connects two very different ideas. The first idea is the *appearance of design*; the second, the *importance of birth and death rates*. By "appearance of design" I mean something that an engineer could look at and surmise that its parts are shaped and arranged so as to carry out some function. Give an optical engineer an eyeball from an unknown species, and the engineer can immediately tell that it is designed for forming an image of the surroundings: It is built like a camera, with a transparent lens, contractible diaphragm, and so on. Moreover, an image-forming device is not just a piece of bric-a-brac, but a tool that is useful for finding food and mates, escaping from enemies, and so on. It is important to note that natural selection explains how the design came to be, using a *second* idea: the actuarial statistics of reproduction in the organism's ancestors. Take a good look at the two separate ideas:

1. A part of an organism appears to have been engineered to enhance its reproduction.
2. That organism's ancestors reproduced more effectively than their competitors.

Note that the two ideas are logically independent. They are about different things: engineering design and birth and death rates. They are about different organisms: The one you are interested in and its ancestors. You can say that an organism has good vision and that good vision should help it reproduce (idea 1), without knowing how well that organism, or any organism, in fact, reproduces (idea 2). Because "design" merely implies an enhanced *probability* of reproduction, all things being equal, a particular organism with well-designed vision may, in fact, not reproduce at all. Maybe it will be struck by lightning. Conversely, it may have a myopic sibling that in fact reproduces better, if, for instance, the same lightning bolt killed a predator who had the sibling in its sights. The theory of natural selection says that idea 2, *the ancestor's birth and death rates*, is the explanation for idea 1, *the organism's engineering design.* Natural selection is not circular in the least.

Chomsky was too flip when he dismissed natural selection as having no substance, as nothing more than a belief that there is some naturalistic explanation for a trait. It is not easy to show that a trait is a product of selection. The trait has to be hereditary. It has to enhance the probability of reproduction of the organism, relative to organisms without the trait, in an environment like the one its ancestors lived in. There has to have been a sufficiently long lineage of similar organisms in the past. And because natural selection has no *foresight*, each intermediate stage in the evolution of an organ is commonly presumed to have conferred some reproductive advantage on its possessor.

Alternatives to Natural Selection

Darwin noted that his theory made strong predictions and could easily be falsified. All it would take is the discovery of a trait that showed signs of design but that appeared somewhere other than at the end of a lineage of replicators that could have used it to help in their replication. One example would be a trait designed only for the beauty of nature, such as a peacock's beautiful but cumbersome tail evolving in moles, whose potential mates are too blind to be attracted to it. Another would be a complex organ that can exist in no useful intermediate form, such as a "part-wing" that could not have been useful for anything until it was 100% of its current size and shape. A third would be an organism that was not produced by an entity that can replicate, such as some insect that spontaneously grew out of rocks, like a crystal. A fourth would be a trait designed to benefit an organism other than the one that created the trait, such as horses evolving saddles. In the comic strip *Li'l Abner*, the cartoonist Al Capp featured selfless organisms called shmoos that laid chocolate cakes instead of eggs and that cheerfully barbecued themselves so that people could enjoy their delicious boneless meat. The discovery of a real-life shmoo would instantly refute Darwin.

Hasty dismissals aside, Chomsky raises a real issue when he brings up alternatives to natural selection. Thoughtful evolutionary theorists since Darwin have been adamant that not every beneficial trait is an adaptation explained by natural selection. When a flying fish leaves the water, it is extremely adaptive for it to reenter the water. But we do not need natural selection to explain this happy event; gravity will do just fine. Other traits, too, need an explanation different from selection. Sometimes a trait is not an adaptation in itself but a consequence of something else that is an adaptation. There is no advantage to our bones being white instead of green, but there is an advantage to our bones being rigid; building them out of the mineral hydroxyapatite is one way to make them rigid, and this mineral happens to be white. Sometimes a trait is constrained by its history, like the S-bend in the human spine that was inherited when four legs became bad and two legs became good. Many traits may be impossible to grow within the constraints of a body plan and the way the genes build the body. The biologist J.B.S. Haldane once said that there are two reasons why humans do not turn into angels: Moral imperfection and a body plan that cannot accommodate both arms and wings. Sometimes a trait comes about by dumb luck. If enough time passes in a small population of organisms, all kinds of coincidences will be preserved in the organisms by a process called **genetic drift**. For example, in a particular generation, all the stripeless organisms might be hit by lightning or die without issue — stripedness will reign thereafter, whatever its advantages or disadvantages.

Stephen Jay Gould and Richard Lewontin (1979) have accused biologists (unfairly, most believe) of ignoring alternative forces and putting too much stock in natural selection. They ridicule such explanations as "just-so stories," an allusion to Kipling's whimsical tales of how various animals got their parts. Gould and Lewontin's essays have been influential in the cognitive sciences, and Chomsky's skepticism that natural selection can explain human language is in the spirit of their critiques.

But Gould and Lewontin's potshots do not provide a useful model of how to reason about the evolution of a complex trait. One of their goals was to undermine theories of human behavior whose political implications they disagreed with. The critiques also reflect their day-to-day professional concerns.

Gould is a paleontologist, and paleontologists study organisms after they have fossilized and become entombed in rocks. Many look more at grand patterns in the history of life than at the workings of an individual's long-defunct organs. When paleontologists discover, for example, that the dinosaurs were extinguished by an asteroid slamming into the Earth and blacking out the Sun, small differences in reproductive advantages understandably seem beside the point.

Lewontin is a geneticist, and geneticists tend to look at the raw code of the genes and their statistical variation in a population rather than the complex organs they build. Adaptation can seem like a minor force to them, just as someone examining the 1's and 0's of a computer program in machine language without knowing what the program does might conclude that the patterns are without design.

The mainstream in modern evolutionary biology is better represented by biologists like George Williams, John Maynard Smith, and Ernst Mayr, who are concerned with the design of whole living organisms. Their consensus is that natural selection has a very special place in evolution and that the existence of alternatives does *not* mean that the explanation of a biological trait is up for grabs and depends only on the taste of the explainer (Dawkins, 1986; Mayr, 1983; Maynard Smith, 1988; Tooby and Cosmides, 1990; Pinker and Bloom, 1990; Dennett, 1983).

Evolutionary Origins of Complex Design

The biologist Richard Dawkins has explained the reasoning of complex design lucidly in his book *The Blind Watchmaker*. Dawkins notes that the fundamental problem of biology is to explain *complex design*. The problem was appreciated well before Darwin. The theologian William Paley wrote:

> In crossing a heath, suppose I pitched my foot against a *stone*, and were asked how the stone came to be there; I might possibly answer, that, for anything I knew to the contrary, it had lain there for ever: nor would it perhaps be very easy to show the absurdity of this answer. But suppose I had found a *watch* upon the ground, and it should be inquired how the watch happened to be in that place; I should hardly think of the answer which I had before given, that for anything I knew, the watch might have always been there.

Paley noted that a watch has a delicate arrangement of tiny gears and springs that function together to indicate the time. Bits of rock do not spontaneously exude metal, which forms itself into gears and springs, which hop into an arrangemeant that keeps time. We are forced to conclude that the watch had an artificer who designed the watch with the goal of timekeeping in mind. But an organ such as an eye is even more complexly and purposefully designed than a watch. The eye has a transparent protective cornea, a focusing lens, a light-sensitive retina at the focal plane of the lens, an iris whose diameter changes with illumination, muscles that move one eye in tandem with the other, and neural circuits that detect edges, color, motion, and depth. It is impossible to make sense of the eye without noting that it appears to have been designed for seeing — if for no other reason than that it displays an uncanny resemblance to the manmade camera. If a watch entails a watchmaker and a camera entails a camera maker, then an eye entails an eyemaker, namely, God.

Biologists today do not disagree with Paley's laying out of the problem. They disagree only with his solution. Darwin is history's most important biologist because he showed how such "organs of extreme perfection and complication" could arise from the purely physical process of natural selection.

And here is the key point. No physical process *other* than natural selection can explain the evolution of a complex organ like the eye. The reason the choice is so stark — God or natural selection — is that structures that can do what the eye does are extremely low-probability arrangements of matter. By an unimaginably large margin, most objects thrown together out of generic stuff, even generic animal stuff, cannot bring an image into focus, modulate incoming light, and detect edges and depth boundaries. The animal stuff in an eye seems to have been assembled with the goal of seeing in mind — but in whose mind, if not God's? How else could the mere *goal* of seeing well *cause* something to see well? The very special power of natural selection is to remove the paradox. What causes eyes to see so well now is that they descended from a long line of ancestors that saw a bit better than their rivals, which allowed them to outreproduce those rivals. The small random improvements in seeing were retained and combined and concentrated over the eons, which led to better and better eyes. The ability of *many* ancestors to see a *bit* better in the *past* causes a *specific* organism to see *extremely* well *now*. Another way of putting it is that natural selection is the only process that can steer a lineage of organisms along the path in the astronomically vast space of possible bodies leading from a body with no eye to a body with a functioning eye.

The alternatives to natural selection can, in contrast, only grope randomly. The odds that the coincidences of genetic drift would result in just the right genes coming together to build a functioning eye are infinitesimally small. Gravity alone may make a flying fish fall into the ocean (a nice big target), but gravity alone cannot make bits of a flying fish embryo fall exactly into place to make a flying fish eye. When one organ develops, a bulge of tissue or some nook or cranny can come along for free, the way an S-bend accompanies an upright spine. But you can bet that such a cranny will not just happen to have a functioning lens and a diaphragm and a retina all perfectly arranged for seeing. It would be like the proverbial hurricane that blows through a junkyard and assembles a Boeing 747. For these reasons, Dawkins argues that natural selection is not only the correct explanation for life on Earth, but is bound to be the correct explanation for anything we might be willing to call "life" anywhere in the universe.

And adaptive complexity, by the way, is also the reason that the evolution of complex organs tends to be slow and gradual. It is not that large mutations and rapid change violate some law of evolution. It is only that complex engineering requires precise arrangements of delicate parts; if the engineering is accomplished by accumulating random changes, the changes had better be small. Complex organs evolve by small steps for the same reason that a watchmaker does not use a sledgehammer and a surgeon does not use a meat cleaver.

What About Language?

So we now know which biological traits to credit to natural selection and which ones to other evolutionary processes. *What about language?* In my mind, the conclusion is inescapable. Everything we have learned about the psychology and neurology of language has underscored the adaptive complexity of the language instinct. It is composed of many parts: syntax, with its discrete combinatorial system building phrase structures; morphology, a second combinatorial system building words; a capacious lexicon; a revamped vocal tract; phonological rules and structures; speech perception; parsing algorithms; learning algorithms.

The many parts are physically realized as intricately structured neural circuits, laid down by a cascade of precisely timed genetic events. What these circuits make possible is an extraordinary gift: the ability to dispatch an infinite number of precisely structured thoughts from head to head by modulating exhaled breath. The gift is obviously useful for reproduction. Anyone can benefit from the strokes of genius, lucky accidents, and trial-and-error wisdom accumulated by anyone else, present or past. Actors can benefit by working in teams, their efforts coordinated by negotiated agreements. Randomly jigger a neural network or mangle a vocal tract, however, and you will not end up with a system with these capabilities. The *language instinct*, like the eye, is an example of what Darwin called "that perfection of structure and co-adaptation which justly excites our admiration," and as such bears the unmistakable stamp of nature's designer, natural selection.

If Chomsky maintains that grammar shows signs of complex design, but is skeptical that natural selection manufactured it, what alternative does he have in mind? What he repeatedly mentions is "physical law." Just as the flying fish is compelled to return to the water and hydroxyapatite bones are compelled to be white, human brains might, for all we know, be compelled to contain circuits for Universal Grammar. As cited in Piatelli-Palmarini (1982), Chomsky says:

> These skills [for example, learning a grammar] may well have arisen as a concomitant of structural properties of the brain that developed for other reasons. Suppose that there was selection for bigger brains, more cortical surface, hemispheric specialization for analytic processing, or many other structural properties that can be imagined. The brain that evolved might well have all sorts of special properties that are not individually selected; there would be no miracle in this, but only the normal workings of evolution. We have no idea, at present, how physical laws apply when 10^{10} neurons are placed in an object the size of a basketball, under the special conditions that arose during human evolution.

We may not know for certain (just as we do not know how physical laws apply under the special conditions of hurricanes sweeping through junkyards), but the possibility that there is an undiscovered corollary of the laws of physics that causes human-sized and human-shaped brains to develop the circuitry for Universal Grammar seems unlikely for many reasons.

At the microscopic level, what set of physical laws could cause a surface molecule guiding an axon along a thicket of associated (glial) cells to cooperate with millions of other such molecules to solder together just the kinds of circuits that would compute something as useful to an intelligent social species as grammatical language? The vast majority of the astronomical ways of wiring together a large neural network would surely create something else: bat sonar, nest-building, go-go dancing, or, most likely of all, random neural noise.

At the level of the whole brain, the remark that there has been selection for bigger brains is, to be sure, common in writings about human evolution (especially from paleoanthropologists). Given that premise, one might naturally think that all kinds of computational abilities might come as a by-product. But if you think about it, you should quickly see that the premise has to have it backwards. Why would evolution ever have selected sheer bigness of brain, that bulbous, metabolically greedy organ? A large-brained creature is sentenced to a life that combines all the disadvantages of balancing a watermelon on a broomstick, running in place in a down jacket, and, for women of childbearing age, passing a large kidney stone every few years. Selection on brain size itself would surely have favored a pinhead! Selection for more powerful computational abilities (language, perception, reasoning, and so on) must have given humans a big brain as a by-product, not the other way around.

But even given a big brain, language does not fall out the way that flying fish fall out of the air. We see language in dwarfs whose heads are much smaller than a basketball (Lenneberg, 1967). We also see it in hydrocephalics whose cerebral hemispheres have been squashed into grotesque shapes or sometimes a thin layer lining the skull like the flesh of a coconut, but who are intellectually and linguistically entirely normal (Lewin, 1980). Conversely, there are "specific language impairment" victims with normal, basketball-sized and -shaped brains and intact analytic processing, some of whom are superb with math and computers. All the evidence suggests that it is the *precise wiring* of the brain's microcircuitry that make language happen, not gross size, shape, or neuron packing. The pitiless laws of physics are unlikely to have done us the favor of hooking up that circuitry so that we could communicate with each other in words.

COULD LANGUAGE HAVE EVOLVED GRADUALLY?

To be fair, there are genuine problems in reconstructing how the language faculty might have evolved by natural selection, though Paul Bloom and I have argued that the problems are all resolvable (Pinker and Bloom, 1990). As P.B. Medawar noted, language could not have begun in the form it supposedly took in the first recorded utterance of the infant Lord Macaulay, who after having been scalded with hot tea, allegedly said to his hostess "Thank you Madam, the agony is sensibly abated." If language evolved gradually, there must have been a sequence of intermediate forms, each useful to its possessor, and this raises several questions.

Three Questions to Answer

First, if language inherently involves another individual, who did the first grammar mutant talk to? One answer might be: the 50% of the brothers and sisters and sons and daughters who shared the new gene by common inheritance. But a more general answer is that the neighbors could have partly understood what the mutant was saying even if they lacked the new-fangled circuitry and just used overall intelligence. Though we cannot parse strings of words such as "skid crash hospital," we can figure out what they probably mean (just as English speakers can often do a reasonably good job of understanding Italian newspaper stories based on similar words and background knowledge).

If a grammar mutant is making important distinctions that can be decoded by others only with uncertainty and great mental effort, it could set up a pressure for them to evolve the matching system that allows the distinctions to be recovered reliably by an automatic, unconscious parsing process. Natural selection can take skills that are acquired with effort and uncertainty and hardwire them into the brain, a phenomenon called the **Baldwin Effect** (Hinton and Nowlan, 1987). Selection could have ratcheted up language abilities by favoring in each generation the speakers that the hearers could best decode and the hearers who could best decode the speakers.

Second, what would an intermediate grammar have looked like? Bates and her collaborators (1991) ask:

> What protoform can we possibly envision that could have given birth to constraints on the extraction of noun phrases from an embedded clause? What could it conceivably mean for an organism to possess half a symbol, or three quarters of a rule? . . . Monadic symbols, absolute rules and modular systems must be acquired as a whole, on a yes-or-no basis — a process that cries out for a Creationist explanation (p. 31).

The question is a bit odd, because it assumes that Darwin literally meant that organs must evolve in successively larger fractions (half, three-quarters, and so on). Bates' rhetorical question is like asking what it could conceivably mean for an organism to possess half a head or three-quarters of an elbow.

Darwin's real claim, of course, is that organs evolve in successively more complex forms. Grammars of intermediate *complexity* are easy to imagine. They would have symbols with a narrower range, rules that are less reliably applied, modules with fewer rules, and so on.

Derek Bickerton (1990) answers Bates more concretely. He gives the term *protolanguage* to chimp signing, pidgins, child language in the two-word stage, and the unsuccessful partial language acquired after the critical period by Genie and other "wolf-children." Bickerton suggests that *Homo erectus* spoke in protolanguage. Obviously there is still a huge gulf between these relatively crude systems and the modern adult language instinct, and here Bickerton makes the jaw-dropping additional suggestion that a single mutation in a single woman, "African Eve," simultaneously wired in syntax, resized and reshaped the skull, and reworked the vocal tract (see also Pinker, 1992).

We can extend the first half of Bickerton's argument without accepting the second half (which is reminiscent of hurricanes assembling jetliners). The languages of children, pidgin speakers, immigrants, tourists, aphasics, telegrams, and headlines all show that there is a vast continuum of viable language systems varying in efficiency and expressive power, exactly what the theory of natural selection requires.

Third, how does each step in the evolution of a language instinct, up to and including the most recent ones, enhance fitness? David Premack (1985) writes:

> I challenge the reader to reconstruct the scenario that would confer selective fitness on recursiveness. Language evolved, it is conjectured, at a time when humans or protohumans were hunting mastodons. . . . Would it be a great advantage for one of our ancestors squatting alongside the embers, to be able to remark: "Beware of the short beast whose front hoof Bob cracked when, having forgotten his own spear back at camp, he got in a glancing blow with the dull spear he borrowed from Jack"?
>
> Human language is an embarrassment for evolutionary theory because it is vastly more powerful than one can account for in terms of selective fitness. A semantic language with simple mapping rules, of a kind one might suppose that the chimpanzee would have, appears to confer all the advantages one normally associates with discussions of mastodon hunting or the like. For discussions of that kind, syntactic classes, structure-dependent rules, recursion and the rest, are overly powerful devices, absurdly so (pp. 281–282).

This objection is a bit like saying that the cheetah is much faster than it has to be, or that the eagle does not need such superb vision, or that the elephant's trunk is an overly powerful device, absurdly so. But it is worth taking this challenge (Burling, 1986; Cosmides and Tooby, 1992; Barkow, 1992; Pinker and Bloom, 1990). Three particularly telling points need to be addressed.

First, bear in mind that selection does not need great advantages. Given the vastness of time, tiny advantages will do. Imagine a mouse that was subjected to a minuscule selection pressure for increased size, say, a 1% reproductive advantage for offspring that were 1% bigger. Simple arithmetic shows that the mouse's descendants would evolve to the size of an elephant in a few thousand generations, an evolutionary eyeblink.

Second, if contemporary hunter-gatherers are any guide, our ancestors were not grunting cave men with little more to talk about than which mastodon to avoid. Hunter-gatherers were accomplished toolmakers and superb amateur biologists with detailed knowledge of the life cycles, ecology, and behavior of the plants and animals they depended on. Language would surely have been useful in anything resembling such a lifestyle. It is possible to imagine a superintelligent solitary species, whose isolated members cleverly negotiated their environment, but what a waste! There is a fantastic payoff in transmitting hard-won knowledge to kin and friends, and language is obviously a major means of doing so.

Grammatical devices designed for communicating precise information about time, space, objects, and who did what to whom are not like the proverbial thermonuclear flyswatter. Recursion in particular is extremely useful, and is not, as Premack (1985) implies, confined to phrases with tortuous syntax. Without recursion one cannot say "the man's hat" or "I think he left." Recall that all you need for **recursion** is an ability to embed a noun phrase inside another noun phrase or a clause within a clause, which falls out of simple rules (for example, in technical notation: "NP → det N PP" and "PP → P NP"). With this ability, a speaker can pick out an object to an arbitrarily fine level of precision.

The abilities can make a big difference. It makes a difference whether a far-off region is reached by taking the trail that is in front of the large tree or the trail that the large tree is in front of. It makes a difference whether the region has animals that you can eat or animals that can eat you. It makes a difference whether the area has fruit that is ripe or fruit that was ripe or fruit that will be ripe. It makes a difference whether you can get there by walking for 3 days or whether you can get there and walk for 3 days.

Third, people everywhere depend on cooperative efforts for survival and form alliances by exchanging information and commitments. This, too, puts complex grammar to good use. It makes a difference whether you understand me as saying that if you give me some of your fruit I will share meat that I will get. Or that you should give me some fruit because I shared meat that I got. Or that if you do not give me some fruit, I will take back the meat that I got. Again, recursion is far from being an absurdly powerful device. Recursion allows sentences such as "He knows that she thinks that he is flirting with Mary" and other means of conveying gossip, an apparently universal human vice.

The Origins of Grammar

But could these exchanges really produce the rococo complexity of human grammar? Perhaps. Evolution often produces spectacular abilities when adversaries get locked into an "arms race," such as the struggle between cheetahs and gazelles. Some anthropologists believe that human brain evolution was propelled more by a **cognitive arms race** among social competitors than by mastery of technology and the physical environment. After all, it doesn't take much brain power to master the ins and outs of a rock or to get the better of a berry. But outwitting and second-guessing an organism of approximately equal mental abilities with nonoverlapping interests, at best, and malevolent intentions, at worst, makes formidable and ever-escalating demands on cognition. A cognitive arms race clearly could propel a linguistic one.

In all cultures, social interactions are mediated by persuasion and argument. How a choice is framed plays a large role in determining which alternative people choose. Thus, there could easily have been selection for any edge in the ability to frame an offer so that it appears to present maximal benefit and minimal cost to the negotiating partner and in the ability to see through such attempts and to formulate persuasive counterproposals.

Anthropologists have noted that tribal chiefs are often both gifted orators and highly polygynous — a splendid prod to any imagination that cannot conceive of how linguistic skills could make a Darwinian difference. I suspect that evolving humans lived in a world in which language was woven into the intrigues of politics, economics, technology, family, sex, and friendship that played key roles in individual reproductive success. They could no more live with a "Me-Tarzan-you-Jane" level of grammar than we could!

FROM SO SIMPLE A BEGINNING, FORMS BEAUTIFUL HAVE EVOLVED

The brouhaha raised by the uniqueness of language has many ironies. The spectacle of humans trying to ennoble animals by forcing them to mimic human forms of communication is one. The pains that have been taken to portray language as innate, complex, and useful, but not as a product of natural selection — the one force in nature that can make innate complex useful things — is another.

Why should language be considered such a big deal? It has allowed humans to spread over the planet and wreak large changes, but is that any more extraordinary than coral that build islands, earthworms that shape the landscape by building soil, or the photosynthesizing cyanobacteria that first released corrosive oxygen into the atmosphere and produced an ecological catastrophe of its time? Why should talking humans be considered any weirder than elephants, penguins, beavers, camels, rattlesnakes, hummingbirds, electric eels, leaf-mimicking insects, giant sequoias, Venus flytraps, echolocating bats, or deep-sea fish with lanterns growing out of their heads? Some creatures have traits unique to their species, others do not, depending only on the accidents of which of their relatives have become extinct.

Darwin emphasized the genealogical connectedness of all living things, but evolution is descent *with modification*, and natural selection has shaped the raw materials of bodies and brains to fit them into countless differentiated niches (see Chapter 4 and Tooby and Cosmides, 1989). For Darwin, such is the "grandeur in this view of life" — "that whilst this planet has gone cycling on according to the fixed law of gravity, from so simple a beginning endless forms most beautiful and wonderful have been, and are being, evolved."

Acknowledgments. This chapter is adapted with permission from Pinker, S. 1994. The big bang. In: *The Language Instinct* (New York: William Morrow), pp. 332–369. Preparation supported in part by the National Institutes of Health Grant HD 18381, National Science Foundation Grant BNS 91-09766, and by the McDonnell-Pew Center for Cognitive Neuroscience at the Massachusetts Institute of Technology.

REFERENCES

Barkow, J.H. 1992. Beneath new culture is old psychology: Gossip and social stratification. In: Barkow, J.H., Cosmides, L., and Tooby, J. (Eds.), *The Adapted Mind: Evolutionary Psychology and the Generation of Culture* (New York: Oxford Univ. Press), pp. 627–637.

Bates, E., Thal, D., and Marchman, V. 1991. Symbols and syntax: A Darwinian approach to language development. In: Krasnegor, N.A., Rumbaugh, D.M., Schiefelbusch, R.L., and Studdert-Kennedy, M. (Eds.), *Biological and Behavioral Determinants of Language Development* (Hillsdale, NJ: Erlbaum), pp. 29–65.

Bickerton, D. 1990. *Language and Species* (Chicago: Univ. Chicago Press), 297 pp.

Brandon, R.N. and Hornstein, N. 1986. From icons to symbols: Some speculations on the origin of language. *Biol. Phil. 1*: 169–189.

Burling, R. 1986. The selective advantage of complex language. *Ethol. Sociobiol. 7*: 1–16.

Caplan, D. 1987. *Neurolinguistics and Linguistic Aphasiology* (New York: Cambridge Univ. Press), 498 pp.

Carrington, R. 1958. *Elephants* (London: Chatto and Windus), 272 pp.

Chomsky, N. 1972. *Language and Mind* (2nd ed.) (New York: Holt, Rinehart, and Winston), 194 pp.

Chomsky, N. 1988. *Language and Problems of Knowledge: The Managua Lectures* (Cambridge: MIT Press), 205 pp.

Corballis, M. 1991. *The Lopsided Ape* (New York: Oxford Univ. Press), 366 pp.

Cosmides, L. and Tooby, J. 1992. Cognitive adaptations for social exchange. In: Barkow, J., Cosmides, L., and Tooby, J. (Eds.), *The Adapted Mind* (New York: Oxford Univ. Press), pp. 163–228.

Darwin, C.R. 1859. *On the Origin of Species*, facsimilie edition, 1964 (Cambridge: Harvard Univ. Press), 502 pp.

Dawkins, R. 1986. *The Blind Watchmaker* (New York: Norton), 332 pp.

Deacon, T.W. 1988. Evolution of human language circuits. In: Jerison, H. and Jerison, I. (Eds.), *Intelligence and Evolutionary Biology* (New York: Springer), pp. 363–382.

Deacon, T.W. 1989. The neural circuitry underlying primate calls and human language. *Hum. Evol. 4*: 367–401.

Dennett, D.C. 1983. Intentional systems in cognitive ethology: The "Panglossian Paradigm" defended. *Behav. Brain Sci. 6*: 343–390.

Galaburda, A.M. and Pandya, D.N. 1982. Role of architectonics and connections in the study of primate brain evolution. In: Armstrong, E. and Falk, D. (Eds.), *Primate Brain Evolution* (New York: Plenum), pp. 203–216.

Gardner, R.A. and Gardner, B.T. 1969. Teaching sign language to a chimpanzee. *Science 165*: 664–672.

Gould, J.L. and Marler, P. 1987. Learning by instinct. *Sci. Am. 256* (January): 74–85.

Gould, S.J. 1985. *The Flamingo's Smile: Reflections in Natural History* (New York: Norton), 476 pp.

Gould, S.J. and Lewontin, R.C. 1979. The spandrels of San Marco and the Panglossian program: A critique of the adaptationist programme. *Proc. R. Soc. London 205*: 281–288.

Hinton, G.E. and Nowlan, S.J. 1987. How learning can guide evolution. *Complex Sys. 1*: 495–502.

Hurford, J.R. 1989. Biological evolution of the Saussurean sign as a component of the language acquisition device. *Lingua 77*: 187–222.

Hurford, J.R. 1991. The evolution of the critical period in language acquisition. *Cognition 40*: 159–201.

Lenneberg, E.H. 1967. *Biological Foundations of Language* (New York: Wiley), 489 pp.

Lewin, R. 1980. Is your brain really necessary? *Science 210*: 1232–1234.

Lieberman, P. 1990. Not invented here. *Behav. Brain Sci. 13*: 741–742.

Maynard Smith, J. 1984. Optimization theory in evolution. In: Sober, E. (Ed.), *Conceptual Issues in Evolutionary Biology* (Cambridge: MIT Press), pp. 289–315.

Maynard Smith, J. 1986. *The Problems of Biology* (Oxford, UK: Oxford Univ. Press), 134 pp.

Maynard Smith, J. 1988. *Games, Sex, and Evolution* (New York: Chapman and Hall), 264 pp.

Mayr, E. 1982. *The Growth of Biological Thought* (Cambridge: Harvard Univ. Press), 974 pp.

Mayr, E. 1983. How to carry out the adaptationist program. *Am. Nat. 121*: 324–334.

Myers, R.E. 1976. Comparative neurology of vocalization and speech: Proof of a dichotomy. In: Harnad, S.R., Steklis, H.S., and Lancaster, J. (Eds.), *Origin and Evolution of Language and Speech. Ann. N.Y. Acad. Sci.* (special vol.) *280*: 745–757.

Newmeyer, F. 1991. Functional explanation in linguistics and the origin of language. *Lang. Commun. 11*: 3–96.

Petitto, L.A. and Seidenberg, M.S. 1979. On the evidence for linguistics abilities in signing apes. *Brain Lang. 8*: 162–183.

Piatelli-Palmarini, M. (Ed.). 1980. *Language and Learning: The Debate Between Jean Piaget and Noam Chomsky* (Cambridge: Harvard Univ. Press), 409 pp.

Pinker, S. 1992. Review of Bickerton's "Language and Species." *Language 68*: 375–382.

Pinker, S. 1994. *The Language Instinct* (New York: William Morrow), 494 pp.

Pinker, S. In Press. Facts about human language relevant to its evolution. In: Changeux, J.-P. (Ed.), *Origins of the Human Brain* (New York: Oxford Univ. Press).

Pinker, S. and Bloom, P. 1990. Natural language and natural selection. *Behav. Brain Sci. 13*: 707–784.

Premack, A.J. and Premack, D. 1972. Teaching language to an ape. *Sci. Am. 227* (October): 92–99.

Premack, D. 1985. "Gavagai!" or the future history of the animal language controversy. *Cognition 19*: 207–296.

Robinson, B.W. 1976. Limbic influences on human speech. In: Harnad, S.R., Steklis, H.S., and Lancaster, J. (Eds.), *Origin and Evolution of Language and Speech. Ann. N.Y. Acad. Sci.* (special vol.) *280*: 761–771.

Sagan, C. and Druyan, A. 1992. *Shadows of Forgotten Ancestors* (New York: Random House), 505 pp.

Savage-Rumbaugh, E.S. 1991. Language learning in the bonobo: How and why they learn. In: Krasnegor, N.A., Rumbaugh, D.M., Schiefelbusch, R.L., and Studdert-Kennedy, M. (Eds.), *Biological and Behavioral Determinants of Language Development* (Hillsdale, NJ: Erlbaum), pp. 209–233.

Seidenberg, M.S. 1986. Evidence from the great apes concerning the biological bases of language. In: Demopoulos, W. and Marras, A. (Eds.), *Language Learning and Concept Acquisition: Foundational Issues* (Norwood, NJ: Ablex), pp. 29–53.

Seidenberg, M.S. and Petitto, L.A. 1979. Signing behavior in apes: A critical review. *Cognition 7*: 177–215.

Seidenberg, M.S. and Petitto, L.A. 1987. Communication, symbolic communication, and language: Comment on Savage-Rumbaugh, McDonald, Sevcik, Hopkins, and Rupert (1986). *J. Exp. Psychol. Gen. 116*: 279–287.

Terrace, H., Sanders, R., Bever, T.G., and Petitto, L.A. 1979. Can an ape create a sentence? *Science 206*: 891–902.

Tooby, J. and Cosmides, L. 1989. Adaptation versus phylogeny: The role of animal psychology in the study of human behavior. *Int. J. Comp. Psychol.* 2: 105–118.

Tooby, J. and Cosmides, L. 1990. On the universality of human nature and the uniqueness of the individual: The role of genetics and adaptation. *J. Pers. 58*: 17–67.

Wallman, J. 1992. *Aping Language* (New York: Cambridge Univ. Press), 191 pp.

Williams, H. 1989. *Sacred Elephant* (New York: Harmony Books), 175 pp.

Wilson, E.O. 1972. Animal communication. *Sci. Am. 227* (September): 52–71.

GLOSSARY

Adaptation. A trait serving a particular function; also, the evolution of such a trait by natural selection for that function.

Adaptationist approach. In evolutionary psychology, a strategy involving the search for adaptive design; adaptations "designed" by natural selection to solve problems of organismal-environmental interaction.

Adaptive problem. In evolutionary psychology, any of numerous problems arising from organismal-environmental interactions, the solution to which results in enhancement or facilitation of reproduction of the organism.

Algorithm. A set of rules (symbolic codes, subroutines, and the like) for approaching a complex problem.

Allometry. The measure and study of the relative growth of a part of an organism in relation to the entire organism.

American Sign Language (ASL). A grammatical system of hand and body movements used for communication by the hearing impaired.

Analogous traits. Morphological or chemical characteristics having a common function shared by members of two or more biological lineages as a result of their convergent or parallel evolution.

Animal Language Research (ALR). A subdiscipline of biology and psychology concerned with communication among animals, especially nonhuman primates.

Animism. A doctrine that the vital principle of biological development is immaterial spirit.

Anthropomorphic. The attribution of human form or personality to an animal or other object.

Audience effect. In Animal Language Research, the effect a group of listeners of the same species (especially of closely related individuals) is perceived as having on the communicating animal.

Austin. A bonobo (pygmy chimpanzee), one of the first nonhuman primates to communicate reliably using a human symbolic system. Studies of his ability to communicate with Sherman, another bonobo, revealed the importance of language comprehension in two-way communication.

Autonomous. Self-governing, self-determined.

Autotroph. Any of various plants and plant-like organisms (photosynthetic protists, cyanobacteria, and photosynthetic bacteria) that are able to use carbon dioxide as their sole carbon source.

Axon. A commonly elongate filamentous structure that protrudes from a nerve cell.

Azimuth. The horizontal component of the position of a celestial body; more specifically, the angular distance of an object along the horizon measured from north toward east.

Baldwin Effect. Incorporation (hardwiring) into the brain of behavioral traits acquired through natural selection.

Barometric pressure. The pressure of the atmosphere.

Bayes' rule. A mathematical equation for computing the probability that a hypothesis is true given specific content-independent observations.

Bonobo. Pygmy chimpanzee.

British Empiricist. Any of various British philosophers, such as David Hume, who theorized that human experience could be summarized in a few measurable principles.

Broca's area. A brain center associated with the motor control of speech.

Central place forager. Any of various animals that start and terminate their foraging excursions always at the same centrally located place, such as social insects that routinely return to their home colony.

Cerebral cortex. The surface layer of gray matter that covers the anterior portion of the brain (the forebrain).

Chimera. An organism or part of an organism composed of two or more genetically distinct cell types.

Circadian. Occurring or functioning in approximately 24-hour periods or cycles (from *circa*, about + *dies*, day).

Cognitive arms race. Competition in mental ability between two or more groups.

Cognitive map. Internal (mental) representation of a topographic map; a mental image of the surroundings that enables a navigator to compute novel routes between known sites.

Content-independent. Rational problem-solving methods that do not embody specific information relevant to different types of problems and therefore operate in the same manner regardless of a given situation or problem content.

Convergent evolution. Development from differing evolutionary orgins in two or more biological lineages of morphological (for example, the wings of insects and birds) or chemical traits (for example, enzymatic proteins) having similar functions.

Darwinism. The evolutionary theory proposed by Charles Darwin that all currently living organisms have evolved from earlier forms in accordance with the principle of natural selection.

Design evidence. A fit between form and function indicating that a particular aspect of an organism's phenotype is an adaptation acquired as a result of natural selection.

Diencephalon. The posterior subdivision of the forebrain.

Domain-general. A mechanism, such as that postulated to govern reasoning, learning, and memory, that is assumed to be content-independent and to therefore operate according to unchanging principles regardless of the topic involved.

Domain-specific. In evolutionary psychology, with reference to certain views and assumptions about the natural world that are designed to operate specifically within a particular area of activity and are thought to be shared by all humans as a result of their common evolutionary derivation.

E-vector. The electric vector of an electromagnetic wave that is perpendicular to both the magnetic vector and the direction of propagation; in optics, the E-vector direction is known as the direction of (linearly) polarized light.

East Atlantic Flyway. A migration route between West Africa and the arctic followed by numerous bird species.

Egocentric. In the context of animal navigation, movement based on a spatial representation in which the (self-centered) system of reference is internal to the organism rather than based on external (Earth-centered) cues.

Eidetic image. A photograph-like, detailed visual image stored in memory.

Electroencephalography (EEG). A technique used to measure electrical activity emanating at characterstic frequencies from the cerebral cortex.

Elevation. The vertical component of the location of an object (for example, a celestial body), usually measured from sea level.

Embryogenesis. The formation and development of the embryo.

***Emx*.** Together with *Otx*, one of two small families of regulatory genes in mammals that appear to follow a temporal and spatial expression pattern in the forebrain that is more or less a mirror image of *Hox* gene expression along the brain stem and spinal cord; *Emx* is named after "empty spiracle," a counterpart gene family in fruit flies.

Entropy. A measure of the disorder within a closed thermodynamic system.

Environment of evolutionary adaptedness (EEA). The ancestral human environment in which selection pressures caused the design of a particular adaptation to be favored over alternatives.

Ephemeris. A mathematical table or function that correlates the azimuthal (horizontal) position of a celestial body (for example, the Sun) with the time of day.

Ethology. The scientific study of animal behavior, especially under natural conditions.

Gamete. A sex cell (sperm or egg) produced via meiosis in animals and via mitosis in plants.

Ganglion. A mass of nerve tissue containing nerve cells external to the brain or spinal cord.

General intelligence. A faculty thought to generate solutions to reasoning problems, postulated to be composed of simple reasoning circuits that are few in number, content-independent, and general-purpose.

Gene. A segment of DNA (deoxyribonucleic acid) involved in the production of a protein or RNA (ribonucleic acid) molecule.

Genetic drift. The tendency, particularly in small populations, for gene frequencies to change by chance, regardless of whether the gene is advantageous.

Genome. The total complement of genetic information-containing DNA in the chromosomes of an organism.

Geocentric. In the context of animal navigation, based on an Earth-related rather than an egocentric (self-centered) system of reference internal to the organism.

Grammar Module. An area of the brain hypothesized to enable humans to learn the rules of grammar.

Grammar. A system of rules that defines the grammatical structure (inflections, syntax, and so forth) of a language.

Great Chain of Being. A doctrine that all species are arrayed in a ladder-like linear hierarchy from less perfect to more perfect with humans at the top.

Great circle. A circle formed by the intersection of a sphere and a plane passing through the center of the sphere; in navigation, the great circle (orthodrome) route represents the shortest distance between two points on the surface of the Earth.

Habituation/dishabituation technique. A method used to study habitual or accustomed activities of animal behavior.

Heredity. The sum of the qualities and potentialities genetically derived from parents.

Heterotroph. Any of various animal and animal-like organisms (fungi, nonphotosynthetic protists, and many types of bacteria) in which organic compounds serve as the principal source of carbon.

Heuristic. Serving to discover or to stimulate investigation.

Homeotic gene. A gene that, when disrupted, produces a homeotic transformation, a malformation in which a segment of the body is transformed into the likeness of some other normal body segment.

Homologous traits. Structures (for example, bones) or molecules (for example, proteins) that are evolutionarily derived from a common ancestor.

***Hox* gene.** Any of numerous genes in vertebrate animals that are homologous to the homeotic genes of fruit flies.

Hunter-gatherer society. A small nomadic group of humans that relies on hunting animals and gathering plants for its food supply.

Instrumental conditioning. A form of learning in which a particular behavior is taught by use of systematic rewards.

Kanzi. A bonobo (pygmy chipanzee) regarded as the first non-human primate to learn a human language without specific training.

Keyboard. An array of colorful symbols used in Animal Language Research.

Knock-out breed. Any of various species or strains in which specific developmental genes have been inactivated experimentally.

Lamarckian. The evolutionary theory proposed by Jean-Baptiste Lamarck postulating the inheritance of environmentally induced adaptive characteristics.

Lana. A bonobo (pygmy chimpanzee) regarded as the first non-human primate to use grammatical sentences for the purpose of communication.

Lateral geniculate nucleus. The visual relay nucleus of the thalamus.

Loxodrome. A curve on a given surface that intersects the meridian at some constant angle; in navigation, a loxodrome course is called a rhumbline.

Mechanist. An adherent of the doctrine of mechanism that holds natural processes to be mechanically determined and capable of complete explanation by the laws of physics and chemistry.

Medial geniculate nucleus. The auditory relay nucleus of the thalamus.

Meiosis. A type of division of the cell nucleus that in animals produces sex cells (sperm and egg).

Mental map. See Cognitive map.
Meridian. A great circle passing through the geographical poles of the Earth.
Midbrain. The middle division of the three primary embryonic divisions of the vertebrate brain.
Mitosis. A type of division of the cell nucleus that results in two daughter cells, each an exact copy of the parental cell.
Module. A standard or replaceable unit.
Motor neuron. Nerve cells directly controlling muscles.
Natural selection. The process of change in gene frequencies in a population caused by differences in fitness; also, differential reproductive success or survival of individuals, organisms, and species.
Neuromere. Any of a front-to-back sequence of neural zones.
Neuron. A nerve cell, the fundamental functional unit of nervous tissue.
Ontogeny. The developmental life history of an organism.
Optic tectum. The major visual analyzer in amphibians, such as frogs; a paired, midbrain, sheet-like structure.
Organogenesis. The formation and development of an organ.
Orthodrome. In navigation, the great-circle route that represents the shortest distance between two points on the surface of the Earth.
***Otx*.** Together with *Emx*, one of two small families of regulatory genes in mammals that appear to follow a temporal and spatial expression pattern in the forebrain that is more or less a mirror image of *Hox* gene expression along the brain stem and spinal cord; *Otx* is named after "orthodentical," a counterpart gene family in fruit flies.
Paired-associate learning. A basic type of learning in which one item in a series serves as a memory trigger, causing recall (as opposed to thinking) of another item.
Panbanisha. A bonobo (pygmy chimpanzee) that was exposed to human language at an earlier age than was her half-sibling, Kanzi, and that has acquired the ability to use a greater variety of abstract symbols than Kanzi.
Paradigm. An example or model.
Paralogous. With reference to gene homologues occurring in the same genome.
Parallel evolution. Development from differing evolutionary orgins in two or more biologic lineages of morphological (for example, the wings of insects and birds) or chemical traits (for example, enzymatic proteins) having similar functions.
Path integration. A method of determining an individual's position relative to the starting point of a journey by continually integrating the linear and angular components of movement, that is, the courses steered and the distances covered.
Phylogenetic approach. The idea that if all organisms are related to one another by common evolutionary descent, similarities should be expected between closely related species such as humans and their closest primate relatives.
Phylogeny. The evolutionary history of a lineage of organisms.
Playback experiments. In studies of animal communication, the use of tape-recorded calls to study responses.
Polarized light. Light in which the electric vector (E-vector) vibrates transversely to the direction of propagation within a single plane (linear polarization) rather than in all directions (as unpolarized light).
Positive allometry. Relative growth such that a part of an organism is proportionately larger in larger individuals or species.
Propositional calculus. A system of logic that allows deduction of true conclusions from true premises, no matter what the subject matter of the premises.
Prosencephalon. The undifferentiated forebrain at early stages of development.
Prosomere. A neuromere of the prosencephalon, the undifferentiated forebrain at early stages of development.
Recursion. A grammatical device in which a noun phrase is embedded within another noun phrase, or a clause within another clause.
Retinotectal. Projection of the retina onto the roof of the midbrain, especially onto that portion known as the superior colliculus.
Retinotopic. Topographic projection of the retinal or visual field onto a brain region.
Rhumbline. See Loxodrome.
Sara. A bonobo (pygmy chimpanzee) regarded as the first nonhuman primate to demonstrate the ability to learn relationships that mimic those expressed in grammar if presented as a series of forced choice-selections.
Search program. In navigation, the systematic reconnaissance of an area around a point on the surface of the Earth at which the search-goal is most likely to be discovered.
Selection pressure. Any condition or process that causes some individuals to leave more descendants than others as a result of differences in genetic makeup.
Semantics. The meaning or relationship of meanings of one or more signs or sounds to represent objects, events, or concepts.
Sensorimotor. Of, relating to, or functioning in both sensory and motor aspects of bodily activity.
Sensory receptor. Any of various cells specialized for sensing aspects of the external environment and other parts of the body.
Sherman. A bonobo (pygmy chimpanzee), one of the first nonhuman primates to communicate reliably using a human symbolic system. Studies of his ability to communicate with Austin, another bonobo, revealed the importance of language comprehension in two-way communication.
Site-based navigation. In the context of animal navigation, the ability of an animal to determine its position using cues at a particular site, irrespective of how it reached the site.
Social contract algorithm. In evolutionary psychology, domain-specific, functionally distinct computational units thought to govern the content of social exchange.
Social insect. Any of various insects, such as bees, ants, wasps, and termites, that live in large colonies that in their most organized form are characterized by sterile worker castes, overlapping generations living together, and division of labor among the members of the colony.
Sociobiology. The comparative study of social organization in animals, including humans, especially with regard to its genetic and evolutionary history.
Standard Social Science Model. The view that the human mind is virtually free of content until influenced by personal experience.
Subcortical. Of, relating to, involving, or being nerve centers below the cerebral cortex.

Subroutine. In computer programming, a sequence of instructions that inform a computer to carry out a set of operations; in neurobiology, a set of rules by which a particular part of the nervous system accomplishes a particular task.

Synapse. The point at which a nervous impulse passes from one neuron to another.

Syntax. A system of rules (grammar) for assembling words into meaningful phrases, clauses, or sentences.

Telencephalon. The anterior subdivision of the forebrain comprising the cerebral hemispheres and associated structures.

Teleology. The fact or character attributed to nature or natural processes of being directed toward an end or shaped by a purpose.

Thalamus. An ovoid mass of brain nuclei providing the major source of inputs to the cortex.

Theory of Mind. The attribution of a mental state (for example, a belief, a desire, or knowledge) to another individual.

Tree of Life. A branching, tree-like representation showing the relatedness of all organisms, commonly based on comparison of the composition of their rRNAs, the ribonucleic acids of their protein-manufacturing ribosomes.

Universal Grammar. An innate, specified set of rules for making certain classes of symbol transformations, commonly regarded as a plausible model explaining the predisposition of humans for language acquisition.

Variation. The differences among individual organisms of a population.

Vector. A physical quantity characterized by a magnitude (linear component) and a direction (angular component).

Vitalism. A doctrine that the functions of a living organism are caused by a vital principle distinct from physicochemical forces.

Washoe. A bonobo (pygmy chimpanzee), the first nonhuman primate to learn to communicate with humans using signs from the vocabulary of American Sign Language for the hearing impaired.

Wason selection task. A test designed to probe the structure of human reasoning in which the subject analyzes violations of a conditional rule having the form *If P then Q*.

Wernicke's area. The posterior part of the superior temporal convolution that houses the auditory word center.

Zygote. A cell formed by fusion (syngamy) of sperm and egg, the earliest formed cell of the embryo of mammals and the spore-producing generation of vascular (land) plants.

SUBJECT INDEX